LESSON IMAGING

in MATH + SCIENCE

Michelle Stephan
David Pugalee
Julie Cline
Chris Cline

LESSON IMAGING in MATH + SCIENCE

Anticipating Student Ideas and Questions
for Deeper STEM Learning

ASCD ALEXANDRIA, VA USA

ASCD®

1703 N. Beauregard St. • Alexandria, VA 22311-1714 USA
Phone: 800-933-2723 or 703-578-9600 • Fax: 703-575-5400
Website: www.ascd.org • E-mail: member@ascd.org
Author guidelines: www.ascd.org/write

Deborah S. Delisle, *Executive Director;* Robert D. Clouse, *Managing Director, Digital Content & Publications;* Stefani Roth, *Publisher;* Genny Ostertag, *Director, Content Acquisitions;* Carol Collins, *Senior Acquisitions Editor;* Julie Houtz, *Director, Book Editing & Production;* Liz Wegner, *Editor;* Georgia Park, *Senior Graphic Designer;* Mike Kalyan, *Director, Production Services;* Valerie Younkin, *Production Designer;* Kelly Marshall, *Senior Production Specialist*

PAPERBACK ISBN: 978-1-4166-2278-9   ASCD product #117008   n10/16

PDF E-BOOK ISBN: 978-1-4166-2284-0; see Books in Print for other formats.

Quantity discounts: 10–49, 10%; 50+, 15%; 1,000+, special discounts (e-mail programteam@ascd.org or call 800-933-2723, ext. 5773, or 703-575-5773). For desk copies, go to www.ascd.org/deskcopy.

Library of Congress Cataloging-in-Publication Data is available for this title.
LCCN: 2016029749

23 22 21 20 19 18 17 16 15 14   1 2 3 4 5 6 7 8 9 10 11 12

LESSON IMAGING in MATH + SCIENCE

Acknowledgments

The authors would like to thank all of the teachers and administrators we have worked with over the last two decades. Without the collaboration of George McManus, Ashley Dickey, Jennifer Smith, Robin Dehlinger, Kristy Bullock, Ericka Allred, and countless other teachers, we could not have written this book. We also owe a debt of gratitude to Carol Collins. If she had not wandered into our lesson imaging presentation in Houston, there would be no book.

Michelle Stephan would like to thank her parents, Don and Nancy Stephan, for providing the type of childhood that encouraged her to dream big. She would also like to thank her family, Chris Robinson and Tai Stephan, for their understanding, patience, and humor throughout the writing of this book.

Chris and Julie Cline would like to thank Michelle Stephan for introducing them to lesson imaging and for the support she has provided the past three years. They would also like to thank Kristy Bullock for supporting their work with Dr. Stephan and for trusting them to be leaders in their school.

Introduction

To me, lesson imaging is a visualization process. You often hear about athletes visualizing their game. Imaging is a process in which teachers visualize what will take place in their classroom when they present a task.

—George McManus, middle school mathematics teacher, Florida

Imagine that your parents or grandparents are celebrating their 50th wedding anniversary this year. You and your siblings want to plan a special event with lots of family and friends, so you decide to throw a party for them three months from now. With the date decided, you need to think through numerous details carefully. Where will the event take place: the city where they currently live? The city in which most of your family lives? Somewhere central to all? Also, what kind of venue will be needed—a formal setting or more of a "party" atmosphere? Whom will you invite? When do invitations need to be sent? Will it be a surprise? And so on.

As the day gets closer, you excitedly begin to imagine the event in your mind, playing it out activity by activity. You imagine where you will seat certain relatives; you know, for example, that Uncle Lee does not get along well with cousin Meagan, so you should place them far away from each other. As you let the image unfold, you realize that there aren't enough non-alcoholic beverages at the party for those who do not drink, such as cousin Christine. Then you remember that even though your parents love pastries, Auntie Donna is on a diet, so you must make sure there are healthy snacks, too. Aunt Connie is allergic to fish, so you must provide an alternative to the salmon you'd planned to serve. And of course Aunt Melanie's family will be late, so folks should probably have about 30 minutes to mingle until the major festivities begin. You schedule the speakers, plan the approximate time frame, gather photos to display in the room, and make the final decisions on the food and drinks the caterers will serve.

You have a clear vision of what the program will look like. You can see your parents' best friend, Julie, speaking, and you imagine the joke you will tell as a nice way to get her off the microphone (she is a bit long-winded). You visualize Uncle Mike taking the microphone and telling that "fish story" again—but he often uses colorful language, so how will you manage what he says? And you can't forget the traditional toast that Todd always makes at family occasions! You visualize all the grandchildren and great-grandchildren running around and dancing to the music.

If only the celebration would unfold as your mind imagines it!

We have all had similar experiences when we plan a party, a bat mitzvah, a wedding, and so on. We plan meticulously for every situation that might occur, given the diversity of people attending the event and interacting with one another in both predictable and unpredictable ways. You know your family traditions, relationships, desires, and motivations, and you can more or less envision how the night will play out. Of course, the event *never* happens the way we imagine it will. Sometimes it goes much better than we thought, and sometimes we wish we had never decided to do it in the first place. Nevertheless, you had an image of how people might act, communicate, and relate to one another. You anticipated a potential time line for events to occur and even imagined a bit of a contingency plan in case it didn't go exactly as you expected.

A very similar thought process goes on in what we call *lesson imaging*. When planning for the next day's class, a teacher might imagine what is going to happen when the bell rings to welcome students into first period science. She will have a warm-up activity on the board, and she imagines students taking the first two minutes to get settled and then getting on with the routine they established during the first two weeks of class. Around five minutes after the tardy bell rings, the teacher anticipates that students will have finished the warm-up activity, and she will have a student explain his or her solution. That shouldn't take more than one minute, and then she can launch into the main exercise for the day. She is really excited about her science lesson because she expects students to be hooked when she shows a short video about the melting polar ice caps. She imagines students getting uncomfortable and a bit worried about the effect of climate change on their environment. The video will take about 15 minutes; she then envisions students writing two reactions to the video and one question in their science journal. After a few students share their reactions, she will state that their work for the next week will center on understanding the effect of human activities on the environment—that one generation's behavior affects the next. The teacher will then launch the major activity for the day. She anticipates how students will engage in a short exploration that simulates the effects of pollution on a small ecosystem. What questions will she ask to help students notice the cause of the dying ecosystem and the effect that pollution has on each part of the whole? What questions might the simulation evoke from students about pollution control? The teacher imagines a prolonged discussion where students raise concerns for the ecosystem and generate ideas about how to become better stewards of the environment. The lesson will end with students writing their ideas about improving pollution control and other actions they can take to preserve ecosystems.

Will the lesson happen as the science teacher imagines? What steps can she take to ensure that the science standards and lesson objectives are met in the student-motivated way she envisions? You might believe that her image will come to fruition if she is passionate enough about science to motivate students. Or you could argue that the catalyst for her image relies solely on the activity she has designed; if it is engaging enough, the lesson will happen just like she imagines. Some teachers might suggest that the right questions will steer the students in this direction. And others will say that a really neat technology program or video will inspire kids to have the deep conversations the teacher imagines.

In this book, we argue that fostering the inspiring, student-driven discussions that support students' deep inquiry into science, technology, engineering, and mathematics (STEM) ideas takes mindful planning and imaging of all the characteristics of a given lesson. A successful STEM unit requires not only displaying passion, choosing engaging activities, asking the right questions, and making effective use of technology; it also necessitates a carefully detailed image of how a chosen activity will play out in the classroom.

Lesson imaging is a term that comes from Alan Schoenfeld's (1998) work on the relationship between teachers' beliefs and goals and the way that teachers expect their plans to unfold in the classroom. In our view, lesson imaging is a pedagogical act during which teachers anticipate the ways in which their planned activities will unfold in interaction with students during real classroom time. It involves a number of practices that go beyond lesson planning. Much like the thought that went into 50th anniversary celebration discussed at the outset, *lesson planning* involves choosing the activities and structuring the time. Lesson imaging goes further by anticipating how students will engage in those activities, the questions they may have, and the questions teachers might ask to promote deeper reasoning about the central goal of the task. In the chapters that follow, we elaborate on our definition of lesson imaging and delve deeply into each component of a lesson image.

Who Is This Book For?

Although teachers can benefit from lesson imaging with any teaching approach, we find it most useful for those teachers interested in learning how to prepare for a more student-centered, inquiry STEM unit.

With direct instruction, the teacher has typically planned a lecture through a PowerPoint presentation or some other text- or lecture-based format. Teachers might ask questions as they provide the facts and explanations, but usually there is little need to anticipate how students are going to engage in the lesson, other than to predict the misconceptions they might have about the information the teacher provides.

With an inquiry approach, however, the teacher presents a problem or laboratory exercise that can provoke students to devise their own ways of solving the task. Hence, the outcome of the lesson is less controllable than if lecture were the main vehicle for learning. With the more open-ended inquiry approach, lesson imaging gives teachers more insight into and control over the intended direction of the lesson, without heavy-handedly pushing the agenda forward with or without the students' understanding.

An example may help here. Mr. Clark, a 7th grade mathematics teacher, poses a problem to his students (Figure 0.1).

FIGURE 0.1

A Task from *Mathematics in Context*

Terry is designing a tile patio. Her design has an orange square in the middle and a white border around it. These patios can be different sizes. Four sizes are shown.

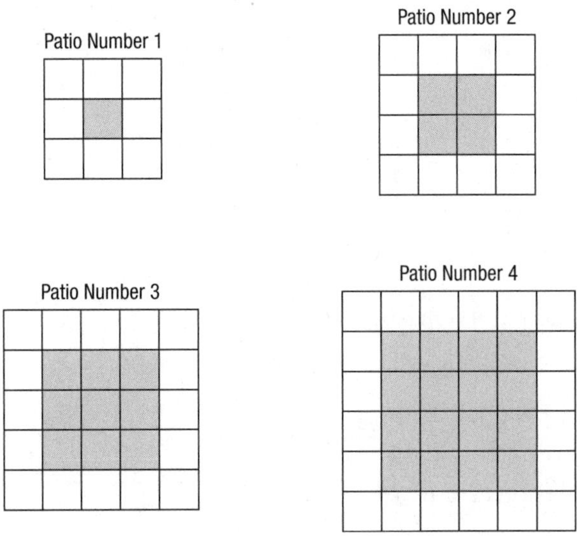

Patio Number 1

Patio Number 2

Patio Number 3

Patio Number 4

Write a direct formula to express the relationship between the total number of tiles (*T*) in any pattern number (*P*).

Source: From *Mathematics in Context* (p. 11), by T. A. Romberg and J. de Lange, 1998, Chicago: Encyclopaedia Britannica. Copyright 1998 by Encyclopaedia Britannica. Reprinted with permission.

Students discuss the problem and share their formulas. Mr. Clark then calls on Tai, who says he came up with $T = P^2 + 4P + 4$.

Mr. Clark writes Tai's solution on the board. He writes P^2 on the orange part of Patio Number 4; circles each of the four white tiles on the top, sides, and bottom of the figure and writes P inside each circle; and then puts an X on the four corners to show the class how Tai created his formula (Figure 0.2).

FIGURE 0.2

Mr. Clark's Symbolization of Tai's Thinking

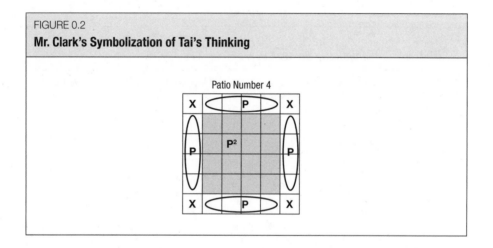

Another student, Tripp, suggests $T = P \times P + 4P + 4$, for very similar reasons.

Mr. Clark is very satisfied with these results and is ready to move on to another problem, when Mary-Riley raises her hand. She suggests another solution: $T = P^2 + (P + 1) \times 4$.

Mr. Clark considers Mary-Riley's solution. Because he had not thought about the problem this way, he has a difficult time, on the spot, determining whether her answer is correct. He tells Mary-Riley that he wants to think about her solution overnight, and then moves on to the next problem.

A teacher who has lesson imaged with colleagues prior to posing this problem might have engaged Mary-Riley and the class differently at this point. Our first question is to you, the reader: is Mary-Riley's solution valid? What does the $P + 1$ stand for in the picture? Try to figure that out before looking at the diagram (Figure 0.3).

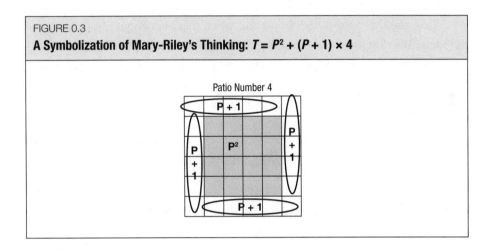

FIGURE 0.3

A Symbolization of Mary-Riley's Thinking: $T = P^2 + (P + 1) \times 4$

The diagram shows that Mary-Riley was structuring the picture into four sets of $P + 1$ tiles plus the P^2 tiles on the inside. Indeed, if one simplifies the expression $P^2 + (P + 1) \times 4$, it is equivalent to both Tai's $P^2 + 4P + 4$ and Tripp's $P \times P + 4P + 4$. Really creative students might even write $T = (P + 2) \times (P + 2)$.

But the question remains: so what?

From a social point of view, capitalizing on Mary-Riley's contribution in the moment might instill a positive mathematical disposition in her and show the rest of the class that there is more than one way to solve the problem.

However, there is another powerful reason to discuss her solution. While one goal of the lesson is to write equations from spatial figures, another mathematical idea that can be explored here is equivalent expressions and equations. Without having to teach simplifying equations in a traditional lecture format, the teacher can ask the class if Mary-Riley's equation is correct or not. Answering that question could be a mini-exploration by students, with the arguments centering on equivalence and simplification.

Lesson imaging includes anticipating as many different solutions as possible. It can help teachers not to be surprised by new solutions, where they must decide in the moment whether a proposed solution would contribute to the mathematical ideas for the lesson.

We argue throughout this book that when teachers engage in the act of lesson imaging prior to instruction, the lesson objectives will be attained more naturally and powerfully from the students than if the lesson were not imaged. Particularly when using an inquiry form of instruction, lesson imaging is crucial so that discussions don't stall and students' ideas are used most powerfully to drive instruction toward important content objectives.

What Are the Components of Lesson Imaging?

Once a worthwhile task has been selected, teachers should imagine how the lesson will unfold by considering the following components:

- Unpacking the lesson objective
- Talking through how to launch the task
- Anticipating how students will engage with the task and what their misconceptions might be
- Deciding which strategies will be presented and in what order
- Deciding what questions to ask to provoke reflection
- Determining what counts as evidence that students have understood the ideas

Almost 10 years ago, we decided to create a lesson imaging template in order to facilitate teachers' discussions that occur as an inquiry activity or scientific exploration is planned. Our template was inspired by the lesson planning of Japanese mathematics teachers (Stevenson & Stigler, 1992), who anticipated both how students will solve the problem the teacher poses and some questions the teacher will use to respond to students' misconceptions. Our template combines both ideas (Figure 0.4).

FIGURE 0.4
Lesson Imaging Template

Science, Technology, Engineering, or Mathematics Goal(s):

State Standard(s):

Cycle 1

Launch (Task presentation)

Exploration (Anticipated student thinking—include class structure [in small groups, with partners, individually] and potential correct and incorrect strategies or solutions)

Whole-Class Discussion (Include tools, symbolizing, technologies, and questions you might pose)

Cycle 2

Launch (Task presentation)

Exploration (Anticipated student thinking—include class structure [in small groups, with partners, individually] and potential correct and incorrect strategies or solutions)

Whole-Class Discussion (Include tools, symbolizing, technologies, and questions you might pose)

Assessment (Evidence of student learning)

This template includes two cycles of exploration, in case teachers are imaging for a block day in which two or more activities can be accomplished. If there is time for a third activity, the teacher can modify the template by adding a Cycle 3. If the class period allows for only one exploration, teachers can skip the second cycle.

The lesson imaging template encourages teachers to begin by unpacking the meaning of the goals they are targeting in the lesson for the day. Are the goals merely procedural skills and practice, or are they more conceptual in nature? For example,

- What does it mean to *understand* photosynthesis?
- What concepts underlie engineering the most stable bridge?

When teachers unpack what it takes for students to understand the content, the activity, exploration, and student discussion will all be more fruitful.

The second major component of the template involves imaging how to launch an exploration that forms the crux of the lesson. The launch is the part of the lesson where teachers pose the problem or experimentation, *not* where they teach students everything they need to know in order to solve the task. Sometimes teachers miss the point of the launch by telling students how to do the exploration rather than letting students explore themselves.

The third and fourth components of lesson imaging entail anticipating students' solutions and then deciding which ones to capitalize on in whole-class discussion. How does the teacher decide which questions to ask to promote powerful ideas in class? What role does symbolizing play in supporting rich discussion? If students actually construct the solution strategies the teacher anticipated, how does the teacher then structure the whole-class discussion so that students share their ideas in a way that builds in an organized fashion and results in the mathematical or science ideas coming to the forefront? The answers to these questions are quite complex and will be explored more deeply to enhance inquiry instruction.

The template ends with an assessment block in order to encourage teachers to write down how they will document student learning each day. Whether through student observation or some type of formal document (e.g., exit slip, homework, quiz), teachers should ensure that some type of assessment is done after each lesson, which will form the basis for lesson imaging for the following class period.

The Catalyst for This Book

Together, we have more than 75 years of experience teaching mathematics and science and working with teachers to help them shift their practice toward an inquiry approach in the STEM fields. Whether as mathematics and science coaches or professional development leaders, we have found that preparing for inquiry-based explorations is best accomplished when teachers work together to lesson image their instruction.

Both Julie Cline and Christopher Cline have worked as science and mathematics teachers for 22 and 21 years, respectively. Because of their success in the classroom, they were asked to coach other teachers in their school regarding lesson imaging in STEM classes. Additionally, they have each presented numerous times in their district about how to support learning communities of teachers who lesson image on a regular basis. They have practiced and promoted lesson imaging firsthand in their own and other teachers' classrooms.

Michelle Stephan was a tenured university professor who left academia to teach middle school mathematics for seven years. During that time, she taught using an inquiry approach and coached teachers in her own school and in schools throughout a large district in Florida. (It was here in 2007 that the idea of lesson imaging began to grow and the lesson image template was formed for use by other teachers throughout the United States.) Stephan is now a tenured faculty member in the College of Education at the University of North Carolina at Charlotte.

David Pugalee is the current director of the Center for STEM Education at the University of North Carolina at Charlotte. He entered higher education after more than a decade teaching middle and high school mathematics and science. His experiences in the classroom raised many questions for him about the role of language and communication in promoting student learning, and he has since worked to improve STEM teaching and learning by focusing on rich experiences that promote students' development of STEM literacy, which is described in more detail in Chapter 1. Lesson imaging is a natural fit, as it provides a tool for exploring instructional planning as a critical component in promoting this vision of student learning.

About This Book

Chapter 1 addresses the meaning of *STEM literacy* and how teachers can effectively incorporate technology and other STEM areas into instruction.

Chapter 2 explores ways to unpack the goals and objectives of a lesson, the resources that teachers can draw on to accomplish this, and how to choose worthwhile problems and explorations that support the lesson objective.

Chapter 3 discusses how to launch an exploration that forms the crux of the lesson.

Chapter 4 looks at how to anticipate students' solutions and then decide which ones to capitalize on in whole-class discussion. How does the teacher decide which questions will promote powerful ideas in class? What role does symbolizing play in supporting rich discussion?

Chapter 5 explores the process of imaging a productive whole-class discussion, following student exploration time.

Chapter 6 brings together all the chapters by "walking through" a full lesson image, with snippets from the actual classroom in which the lesson was taught.

The final chapter examines how teachers can start this process, how STEM mentors and coaches can help teachers engage in lesson imaging, the role that administrators can play in supporting their teachers, and what resources are needed from administrators.

Before Reading Chapter 1 . . .

Consider these questions before moving on to the next chapter:

- How do *you* define STEM literacy?

• What are the important components to consider in a STEM lesson?

• What dispositions do you believe are important for your students to develop in STEM classes?

1

STEM Literacy:
The Nature of STEM Teaching and Learning

Mathematics in the work place makes sophisticated use of elementary mathematics rather than, as in the classroom, elementary use of sophisticated mathematics.

—Lynn Arthur Steen, *Quantitative Literacy*

Before getting into lesson imaging and its implementation in STEM classrooms, let's take a moment to think about the desired outcomes of a standards-based STEM program with a strong inquiry instructional model. STEM has become a buzzword used by many in hopes of capturing the synergy behind the demand for qualified workers in science, technology, engineering, and mathematics, and many schools have thus embraced the term to describe their programs or their curricular emphasis. STEM magnets and charter schools are popping up with great frequency, attempting to capitalize on the national trend and the increased funding. Adopting a tag, however, doesn't necessarily mean that schools have significantly changed their practices or curriculum in the ways necessary to prepare students for college-level STEM studies or the technical entry-level STEM job market. A meaningful view of what STEM education means is central to developing ideas about effective teaching and learning. This chapter provides one way of conceptualizing STEM education, with the intent of establishing some common perspectives that will guide the development of strategies for effective lesson imaging and teaching.

The Four Pillars of Learning

UNESCO's four pillars of learning (Nan-Zhao, 2008; Zollman, 2012) provide a useful framework from which to develop powerful ideas about STEM teaching and learning:

1. Learning to know
2. Learning to do
3. Learning to live together
4. Learning to be

These four pillars promote a continuum on which STEM literacy can be characterized, and they will move us toward a common vision of what it means to have STEM literacy.

Learning to Know

The first pillar, *learning to know*, involves increasing students' literacy in each of the four content areas: science, technology, engineering, and mathematics. These content area literacies are central to the development of lesson imaging as a planning tool to promote effective instruction. As with most publicly popular terms, definitions of content literacy are so diverse that it is hard to pinpoint just one. Regardless, being literate in each of the four content domains serves as the crux of lesson imaging for STEM education, and we will thus define what these literacies mean for us in the context of this book.

Scientific literacy involves constructing the content and process skills necessary to understand the natural world. Beyond having a conceptual understanding of the world, being scientifically literate also means being able to "use the methods of science; apply science to social, economic, political, and personal issues; and develop an appreciation of science as a human endeavor and intellectual achievement" (Hurd, 1958, p. 13). The most important aspects of scientific literacy involve knowing the content and practices of science well enough to make informed decisions about the natural world around us. For example, individuals who are scientifically literate can understand both positive and negative implications of building a nuclear facility in their town and can make reasoned, factual arguments for and against such a proposal.

Technological literacy goes beyond the ability to simply use digital devices—it is the "ability to use, manage, assess and understand technology" (International Technology Education Association, 2007, p. 7). Being technologically literate involves using the scientific method employed by engineers and scientists to create new technologies and being able to assess both the value of a technology and the potential harm it might create—in other words, determining whether a technology is worth pursuing.

Engineering literacy involves knowledge of and facility with the design method that is employed in creating and testing new innovations and understanding the implications of such products.

Mathematical literacy refers to the "capacity of students to analyze, reason and communicate effectively as they pose, solve and interpret mathematical problems in a variety of situations involving quantitative, spatial, probabilistic or other mathematical concepts" (Organisation for Economic Co-operation and Development, 2007, p. 304).

It is clear that STEM literacy includes knowing the content of the discipline at more than a rote level, being able to employ the scientific method or engineering design process when exploring a domain or designing new tools, and assessing and communicating the impact of any findings on the natural world.

Learning to Do

Teaching and learning in STEM extend beyond an emphasis on memorizing content. STEM literacy also involves *learning to do*, the second pillar of learning. Employing inquiry-guided instructional methods gets students involved in ways that incorporate the higher cognitive skills that are indicative of 21st century learning. Exploring innovative problems provides interesting and challenging opportunities for students to develop problem solving, optimization, and visualization in mathematics, science, and engineering contexts (Binkley et al., 2012).

The emphasis on 21st century skills in the United States and on those skills necessary to better understand the global environment, as envisioned by the European Commission (Delors, 2013), requires teachers to move beyond a restrictive view of skill development to one that moves learners to be self-confident in their learning and capable of dealing with life's challenges, both professional and personal. STEM teaching and learning should thus involve active engagement, where students *learn by doing* in ways that involve setting goals, formulating hypotheses, and predicting outcomes as students organize, prioritize, research, formulate, test, and verify their ideas.

Learning to Live Together

Ideally, a critical outcome of effective teaching would be *learning to live together*, the third pillar of learning. As an ancient Chinese proverb says, "Tell me and I forget, show me and I remember, involve me and I understand." Unfortunately, communication and team collaboration skills are generally not considered in the critical instructional planning stage.

When students are involved in sustained engagement, they also take part in collaborative inquiry. This advances a shared knowledge and facilitates the development of *meta-skills*, higher-level thinking processes that emerge from sustained engagement and collaboration (Binkley et al., 2012). These skills are critical to inquiry teaching.

Collaborative skills do not develop in isolation. Effective teaching involves making deliberate choices about how collaboration will happen in the classroom. In the long term, the development of these skills promotes the type of collaboration and cooperation that increases a sense of community.

Learning to Be

A well-organized inquiry environment promotes the type of STEM learning that results in students' development of self-regulation and self-determination. In other words, students develop the cognitive, affective, and psychomotor skills that are part of the lifelong learning process—they are *learning to be*. Puttnam (2015) challenges teachers to support the development of "educational assets" that allow youth to "solve problems, tackle challenges, work in teams, and learn how to communicate" (pp. 122–123). This kind of learning leads to autonomous and fulfilled learners—a hallmark of inquiry teaching.

We may think of this type of autonomy as a *productive struggle*, one that fosters understanding, encourages setting goals that are attainable and worthwhile, and gives students a sense of empowerment. According to Warshauer (2015), instructional approaches that consider students' struggles and support and guide their thinking toward a productive resolution strengthen students' disposition toward tackling challenging tasks, ultimately leading to persistence and understanding.

What Does This Look Like?

The vision of STEM literacy exemplified through UNESCO's four pillars of learning raises a clear challenge: teaching has to be different in order to accomplish such powerful outcomes.

A New Standards Vision

The Common Core State Standards and the Next Generation Science Standards both provide a context for revisualizing how knowledge and understanding are constructed. Figure 1.1 illustrates the synergy among these process skills (Cheuk, 2012).

Keep in mind that the learning outcomes desired from strong, inquiry-focused instruction and vibrant, connected instruction will be evident. Instruction will build on literacy and mathematics by calling for earlier and more frequent work with informational texts, writing with an emphasis on analysis and presentation, the construction of viable arguments, and critique of the reasoning of others. Modeling, which is emphasized in the secondary grades, involves analysis and decision making—validating conclusions through comparisons with the situation or problem context and then improving the model or reporting on one's conclusions and reasoning. This emphasizes the choices, assumptions, and approximations that are present in the cycle (Stage, Asturias, Cheuk, Daro, & Hampton, 2013).

FIGURE 1.1

Commonalities Among Science, Mathematics, and English Language Arts

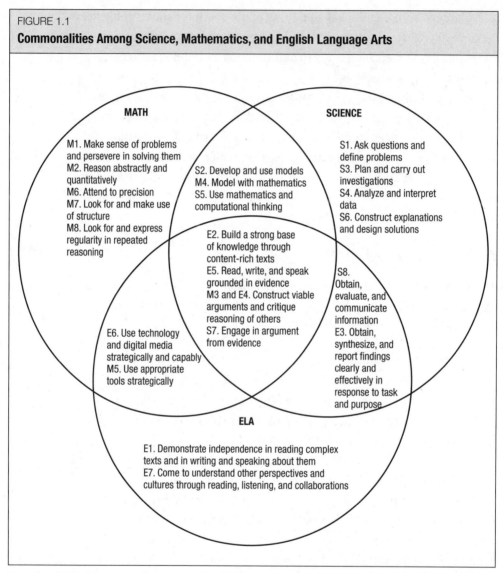

MATH

M1. Make sense of problems and persevere in solving them
M2. Reason abstractly and quantitatively
M6. Attend to precision
M7. Look for and make use of structure
M8. Look for and express regularity in repeated reasoning

S2. Develop and use models
M4. Model with mathematics
S5. Use mathematics and computational thinking

SCIENCE

S1. Ask questions and define problems
S3. Plan and carry out investigations
S4. Analyze and interpret data
S6. Construct explanations and design solutions

E2. Build a strong base of knowledge through content-rich texts
E5. Read, write, and speak grounded in evidence
M3 and E4. Construct viable arguments and critique reasoning of others
S7. Engage in argument from evidence

S8. Obtain, evaluate, and communicate information
E3. Obtain, synthesize, and report findings clearly and effectively in response to task and purpose

E6. Use technology and digital media strategically and capably
M5. Use appropriate tools strategically

ELA

E1. Demonstrate independence in reading complex texts and in writing and speaking about them
E7. Come to understand other perspectives and cultures through reading, listening, and collaborations

Source: From "Relationships and Convergences Among the Mathematics, Science, and ELA Practices," by T. Cheuk, 2012, Palo Alto, CA: Stanford University. Copyright 2012 by Tina Cheuk. Reprinted with permission.

One way of thinking about the connection between STEM literacy and the mathematics, science, and engineering standards is the idea of *operationalized inquiry* (Stage et al., 2013), which is related to the eight practices of science and engineering (Committee on Standards for K–12 Engineering Education & National Research Council, 2010):

1. Asking questions and defining problems
2. Developing and using models
3. Planning and carrying out investigations
4. Analyzing and interpreting data
5. Using mathematics and computational thinking
6. Constructing explanations and designing solutions
7. Engaging in argument from evidence
8. Obtaining, evaluating, and communicating information

The standards present the profession with thinking and cognitive processes that challenge educators to consider both their teaching practice and their lesson content, promoting a model of teaching and learning that espouses a broader and richer view of what it means to be literate in STEM.

A Dynamic Teaching Process

Lesson imaging provides a powerful tool for teachers to consider the dynamic and multifaceted nature of the instructional process effectively. For teachers to anticipate how their plans will unfold *and* how students will engage in the lesson, an understanding of the nature of STEM literacy is important. Lessons that promote the outcomes envisioned in this discussion include *content knowledge*, *discursive processes*, and *literacy skills*. Figure 1.2 captures the complexity of the relationship among these three outcomes; the arrows show how each component connects with the other two. Lesson imaging provides opportunities to focus on these connections during planning, with later follow-up and reflection.

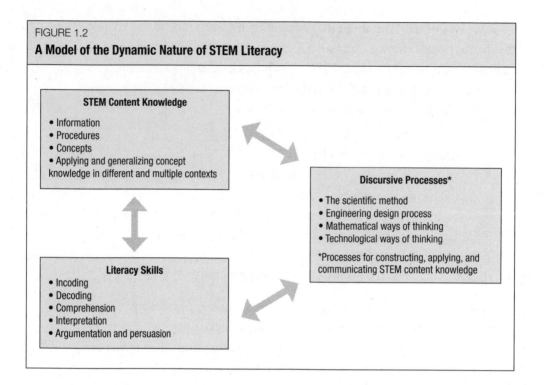

FIGURE 1.2

A Model of the Dynamic Nature of STEM Literacy

STEM Content Knowledge

- Information
- Procedures
- Concepts
- Applying and generalizing concept knowledge in different and multiple contexts

Discursive Processes*

- The scientific method
- Engineering design process
- Mathematical ways of thinking
- Technological ways of thinking

*Processes for constructing, applying, and communicating STEM content knowledge

Literacy Skills

- Incoding
- Decoding
- Comprehension
- Interpretation
- Argumentation and persuasion

STEM with Caution

As discussed, STEM literacy involves learning *content knowledge* from each of the overlapping literacy domains: scientific, technological, engineering, and mathematical. Because STEM education is a relatively new research direction, it is unclear whether it is best to teach an integration of all four disciplines at once or to focus on each domain individually. What is important to note is that current conceptions of content knowledge go beyond "compartmentalized" knowledge in any one of these domains; instead, teachers must begin to think more about the interconnections among these disciplines.

It is rare to find a 60-minute lesson that emphasizes each STEM domain equally. Many STEM lessons we have seen tend to favor one content area (e.g., science) at the expense of the others, which are then taught rotely to the students in order to finish, say, a science lab. Rather than force-fit all four content areas, teachers might think about the connections that are evident and which domains are primary. It is critical to build connections that are clear and natural and to avoid forced linkages that do little to build students' understanding of the interrelationships among the disciplines. Considering the information, procedures, and concepts that can be developed naturally through a lesson is the foundation for providing a context where students apply and generalize concept knowledge in different and multiple contexts, as depicted in the model seen in Figure 1.2.

In the model, *discursive processes* refers to the idea that conclusions are constructed through reason and that thinking is characterized by analytic reasoning. These statements capture the type of discourse that will make a difference in the learning experienced by students. For example, consider the nature of "doing" science. Engaging students in scientific practices allows them to develop scientific knowledge in meaningful contexts that resemble how actual scientific discoveries are made (Evagorou, Erduran, & Mäntylä, 2015). The scientific methods, engineering design processes, mathematical practices, and technological ways of thinking and acting all involve discursive processes inherent in the disciplines and position the learner in dynamic interplays that involve students in thinking deeply about how "work" progresses for professionals.

It can be argued that the discursive processes supported by inquiry teaching hold the same power for all STEM fields. Inquiry engages the learner in "doing" STEM and supports the processes by which professionals in the various fields engage in inquiry, make discoveries, and build knowledge. Vital to the deep learning of content and concepts via discursive practices is the uncovering of how those processes play out in the classroom and how they support students' engagement and learning. Lesson imaging affords the opportunity to look inward at teachers' perceptions and intent, as well as to look outward at the actual reality of classrooms.

Literacy skills in the STEM disciplines should also be considered. Students need to become aware of literacy tools and processes that will assist them in both understanding text and producing text that effectively conveys information and concepts. Literacy tools must be an integral part of the instructional process. Teachers should use—and model the use of—these tools to build student competence and to promote development of high-level reasoning and effective communication. Lesson imaging provides a tool for teachers to consider the use of literacy tools in unpacking a lesson: how to structure student experimentation and exploration; how to support clear, concise, and effective discussion; and how to foster other types of communication, including both spoken and written forms. Supporting literacy development in the context of STEM will further students' ability to develop the skills necessary to encode, decode, comprehend, interpret, and argue effectively within the technical language contexts characteristic of the STEM fields.

An example of a lesson that captures many of the STEM literacy ideas outlined above is Packages and Polygons, a *Mathematics in Context* unit (Romberg & de Lange, 1998), which integrates two- and three-dimensional spatial relationships and volume. In one activity, students use straws and clay to build "bar models" of a cube and a triangular prism (Figure 1.3). Students then determine which polyhedron is the most stable, and they can add additional straws to the structure to improve its stability. This lesson engages students in the engineering process of designing, testing, and revising their products to determine which shape is the most stable. Students make arguments based on evidence collected during trials, and they compare and contrast their conclusions with their classmates in order to determine why certain structures are more stable than others.

As students present their findings orally, they learn a wealth of mathematical vocabulary words, including *vertex*, *diagonal*, *edge*, *polyhedron*, and *faces*. Additionally, this activity is placed carefully in a series of lessons that lead to measuring the volume of three-dimensional shapes.

As can be seen, not all content areas are used in this STEM activity; both science and technology are minimized. However, one might argue that the created shapes are in fact a technology because they are made by humans. In addition, these two domains may become primary in future lessons within this unit. What is key is that this lesson evokes the design process and both critical thinking and problem solving, which form the foundation of all STEM literacy. Hence, it may not be practical or desirable to have all four STEM content areas in each lesson; instead, they can coexist in the larger unit, with the engineering design principles and critical inquiry as the basis of each lesson.

FIGURE 1.3

A Problem from Packages and Polygons

Which bar model is more stable: a triangular prism or a cube? Why do you think so?

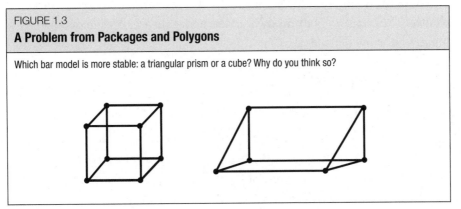

Source: From *Mathematics in Context* (Packages and Polygons, p. 17), by T. A. Romberg and J. de Lange, 1998, Chicago: Encyclopaedia Britannica. Copyright 1998 by Encyclopaedia Britannica. Reprinted with permission.

Conclusion

This chapter provides some thinking about what the "big picture" of STEM literacy should look like for students. The ideas presented challenge us to think differently about the outcomes of teaching and to make carefully planned instructional decisions. Only through such efforts can STEM literacy, as envisioned in this chapter, be achieved.

The discussion reminds us of a speech given by a top executive of a well-respected high-tech company. He lamented that what was lacking in today's graduates was *innovation*. The problem with the speech, however, was that the executive really didn't provide any ideas about what he meant by *innovation*. It has become an empty buzzword for too many leaders who simply want to lament the current state of education.

Perhaps the vision presented in this chapter is the answer to what it means to develop students who are innovators: those who can collaborate with others to develop questions based on observations or evidence, design and test ideas, formulate conclusions, identify outcomes, and engage in thinking and communicating in deep ways that promote argumentation and the generation of ideas.

With this vision in mind and lesson imaging as a tool, instruction *will* change. The result will be empowered learners who are capable of responding to tomorrow's challenges.

Before Reading Chapter 2...

Consider these questions before moving on to the next chapter:

- How is STEM literacy currently integrated into your own instruction?
- Where do you find resources to develop STEM units and lessons?
- What does it mean to *understand* a STEM idea such as ratio? Mitosis? Stability?

Beginning the Imaging Process:
Unpacking the Goals

One challenge faced when trying to understand the mathematics in the lesson is that a teacher (experienced or not) naturally understands the mathematics with a "teacher brain." I had to "forget" what I knew about the mathematics as a teacher and try to predict how students in my classroom would respond. I needed to think like my students who were seeing the goal of the lesson for the first time and constantly be listening to student ideas about the mathematics.

—Ashley Dickey, middle school mathematics teacher, Florida

The purpose of this chapter is to elaborate on the lesson imaging process, which begins with understanding the goals of the lesson. Recall that the focus of this book is on *lesson* imaging, not unit or course development. For designing weeklong units on particular STEM concepts, we refer the reader to *Understanding by Design* (Wiggins & McTighe, 2005), which provides extensive information on the backward design process involved in creating units of instruction. However, in order to understand lesson imaging—in particular, unpacking the goals of a lesson—it is necessary to make a few comments about the process of finding and selecting instructional units that are conducive to STEM teaching.

The lessons that are implemented in the classroom must be adapted from a coherent unit that builds students' conceptual understanding in an organized and sequential manner. Teachers must have an idea of how the lesson goals fit within the broader unit. Planning a vacation serves as a nice metaphor, in that the teacher must know where the journey begins and where it leads, not to mention how the current lesson fits within the larger trip.

Finding strong STEM units that focus on inquiry can be a daunting task for teachers, especially those who are new to the approach. In the next few sections, we elaborate on some characteristics that teachers should consider when designing or choosing units for the STEM classroom.

The Coprinciples of STEM Units: Modeling and the Inquiry Method

There are different opinions regarding the best ways to integrate STEM into the classroom. For example, some researchers and educators argue that science, technology, engineering, and mathematics should be included in every lesson within a mathematics classroom. However, this is a formidable, if not impossible, task, especially in light of the perspective held by some that the four disciplines really have very little in common (Clarke, 2014), and often one or more of the disciplines will suffer at the expense of the other. In contrast, others suggest that STEM concepts *should be* interdisciplinary and not taught in isolation of one another. While research on this topic is relatively new, it suggests that there are numerous reasons to integrate the disciplines, though not at the expense of any one discipline itself. Perhaps this is best accomplished by designing STEM units or STEM programs, rather than integrating STEM into every lesson. This approach is especially plausible given the definitions of science, mathematical, technological, and engineering literacy, each of which emphasizes the need for manipulating available resources to solve problems through analyzing, modeling, testing, and revising (English, 2015). In this view, the activity of creating viable models as a solution to a genuine, realistic problem comes to the fore of students' daily educational experiences.

Given these different opinions, we currently advocate the latter approach of designing STEM *programs* and, when feasible, STEM *units* within the school, rather than attempting to insert STEM into each lesson that is taught.

What might a STEM program look like that takes into account the nature and intended outcomes for STEM described in Chapter 1? As you develop your STEM program, keep in mind the four pillars of learning that should form the foundation of STEM programs: learning to know, learning to do, learning to live together, and learning to be (Nan-Zhao, 2008; Zollman, 2012). The main courses that already exist in schools, such as mathematics, science, social studies, and language arts, would continue to exist, as well as other important classes, such as art, physical education, chorus, and band, to name a few. These courses would integrate multiple disciplines into their units as feasible (e.g., the bar model lesson from Chapter 1, which integrates mathematics and engineering). Other courses that are core to the STEM disciplines should then be added, such as robotics, various engineering courses (e.g., aerodynamics, thermodynamics, aerospace), digital literacy, and computer programming. In addition to adding more STEM options, mathematics and science classes should be revised to include *modeling* and *the inquiry method*, the processes that are inherent in each of the STEM disciplines and serve as the conceptual glue of the four content areas.

Modeling

Modeling is the activity of making observations about a given, problematic situation and creating a representation that exhibits the behavior or results of that situation and also allows us to make predictions about the situation. As Quarteroni (2009) argues, mathematical modeling has become increasingly popular in many fields and industries. This is due to the fact that large-scale data analysis and computation can be conducted by computers, thus making real-world modeling a paramount educational objective. Therefore, each STEM discipline needs to incorporate the practices of modeling in its course goals.

The Inquiry Process

A second general practice that coexists within the four disciplines is the use of the inquiry method to either study a problem or design a scientific or technological innovation. We use the term inquiry to include the scientific method (NGSS Lead States, 2013), the Standards for Mathematical Practices and Problem Solving Techniques (National Governors Association Center for Best Practices & Council of Chief State School Officers, 2010; Polya, 1957), the engineering design process (Committee on Standards for K–12 Engineering Education & National Research Council, 2010), and the Standards for Technological Literacy (International Technology Education Association, 2007). Whether engineering a new technology, attempting to understand a scientific phenomenon, or justifying a mathematical idea, all four STEM domains involve the human activity of inquiry, with slight variations depending on the discipline. The inquiry process can be thought of as the systematic process of forming a question, creating a model, experimenting or designing, collecting and analyzing data, and constructing explanations of the phenomenon studied (or innovation designed) to communicate solutions or understandings to the larger community.

The inquiry process is so important to STEM subject areas that the International Technology Education Association devotes three standards to this idea (Figure 2.1).

FIGURE 2.1

From the International Technology Education Association Standards for Technological Literacy

Design

Standard 8: Students will develop an understanding of the attributes of design.

Standard 9: Students will develop an understanding of engineering design.

Standard 10: Students will develop an understanding of the role of troubleshooting, research and development, invention and innovation, and experimentation in problem solving.

Source: From *Standards for Technological Literacy: Content for the Study of Technology* (3rd ed.), by the International Technology Education Association, 2007, Reston, VA: Author. Copyright 2007 by the International Technology Education Association.

Incorporating Modeling and the Inquiry Method

Just as scientists, engineers, and mathematicians use the inquiry method to create models in problem solving, educators can do so as well. For example, consider a task in which middle school teachers challenge their students to design a gift for all teachers in their school for Teacher Appreciation Day. The students must outline the design criteria and constraints, including that the cost of each gift must be below $1.50, the gift must be made in the school's technology laboratory, and the gift should be useful to the teachers. The students work in teams to brainstorm a design to be shared with the entire class. The class decides which gift they will create, and students return to their teams to research different materials that could be used for the gift and to create a model, which they again share with their classmates. Students evaluate each model and then make their final decision (International Technology Education Association, 2007, p. 96).

Engaging in this type of open-ended modeling activity incorporates problem solving *with a realistic purpose:* creating a model for a product that has an intended use. It can be viewed as authentic inquiry that is open-ended and yet structured by the design process.

Characteristics of STEM Programs and Instructional Activities

Given the importance of modeling and the inquiry method, how do teachers or curriculum specialists find instructional materials for their science and mathematics classrooms? In this section, we offer some characteristics that should be considered when vetting instructional units for the classroom and some organizations that can aid in identifying potential STEM resources. Whether designing your own materials or adapting existing instructional units, there are several important factors to consider:

- Does the unit prompt students to model realistic problem situations?
- Is inquiry the essential activity of the unit? (If the lessons begin with instructions on how to solve problems and then show examples before students have a chance to problem solve, the unit does not follow the inquiry approach.)
- Are the lessons within a unit sequenced so that they begin with modeling concrete situations and move toward more abstract models (i.e., there is flow from one lesson to the next, rather than topics presented in isolation)?

• Are there opportunities for genuine STEM discourse (e.g., students must communicate their solution methods and justify their reasoning)?

• Are there opportunities for students to use technologies[1] or create new ones?

• Do students have occasions to work with multiple representations?

• Are students required to communicate their findings in oral and written texts?

• Do the tasks involve high cognitive demand?

These criteria are also consistent with the Strands of Mathematical Proficiency outlined in *Adding It Up* by the Mathematics Learning Study Committee (2001) (Figure 2.2).

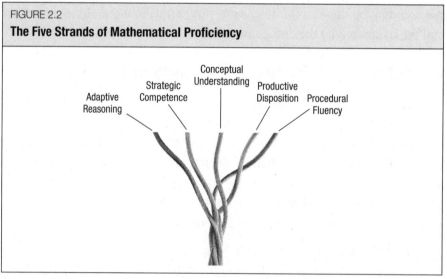

FIGURE 2.2

The Five Strands of Mathematical Proficiency

Source: From *Adding It Up: Helping Children Learn Mathematics* (p. 117), by the Mathematics Learning Study Committee, 2001, Washington, DC: The National Academies Press. Copyright 2001 by the National Academy of Sciences. Reprinted with permission.

---

[1] *Technology* is meant to refer to all devices that are human-made, not solely to digital technologies such as tablets, computers, phones, and apps. Consult the Standards for Technological Literacy (International Technology Education Association, 2007) for a more thorough discussion of what constitutes technology.

Although the Strands are specific to mathematics, it is easy to imagine their crossover to the other STEM disciplines. The Strands include five types of reasoning that are intertwined to form full proficiency in the discipline. Textbooks and instructional materials in the STEM area should focus on developing students' procedural and factual knowledge in the domain (Procedural Fluency Strand). Furthermore, the instructional activities should be constructed to engage students in inquiry that develops their conceptual understanding of key concepts in the discipline, as guided by state standards or the Common Core standards (Conceptual Understanding Strand).

We group the other three Strands together, as they are more general ways of engaging in STEM classes and relate most specifically to the Standards for Practice in the Common Core State Standards for Mathematics (National Governors Association Center for Best Practices & Council of Chief State School Officers, 2010) and the National Council of Teachers of Mathematics (2000) Process Standards:

• The Strategic Competence Strand refers to the ability to solve problems, essentially, by modeling them.

• The Adaptive Reasoning Strand involves the ability to make viable arguments and justify one's reasoning when problem solving.

• The Productive Disposition Strand means developing an attitude toward problem solving as something useful, meaningful, and sensible while also fostering *intellectual autonomy*—a belief that we ourselves can be the problem-solving authority rather than depend on the teacher (Kamii, 1982). (This concept is explored in more detail later in this chapter.)

Let's look at an example adapted from a textbook, which illustrates a typical introduction to solving linear equations—one that does *not* fit within the STEM philosophy (Figure 2.3).

In this example, the page begins with conventional definitions and then shows students a set of steps to follow in order to solve an equation using subtraction. In the left-hand margin, you can even see the authors' ready-made model, which students should use to understand the method. You can imagine how the textbook continues, showing how to solve equations with

FIGURE 2.3

Page Adapted from a Traditional Textbook

To solve an equation, you must **get** the variable by itself. You do this by getting the variable alone on one side of the equation.

Use the properties of equality and **inverse operations** to get the variable alone. An **inverse operation** undoes another operation. For example, subtraction is the inverse of addition. When you solve an equation, each inverse operation you perform should produce a simpler equivalent equation.

PROBLEM 1.1—Solving Equations Using Subtraction

What is the solution of $x + 13 = 27$?

How can you visualize the equation?

You can *draw* a *diagram*. Use a model like the one below to help you.

27

| x | 13 |

THINK

You need to get *x* by itself. Start by writing the equation.

Undo addition by subtracting the same number from each side.

Simplify each side of the equation.

Check your answer.

WRITE

$x + 13 = 27$

$x + 13 - 13 = 27 - 13$

$x = 14$

$x + 13 = 27$
$14 + 13 = 27$

ALERT!!

Remember, what you do to one side, you must do to the other side.

addition and the other operations. There is no opportunity for students to create models of their activity or to problem solve—the *problems have been solved for them.*

In contrast, consider one of the beginning activities on writing equations from *Mathematics in Context* (Romberg & de Lange, 1998), a textbook that is meant to be used with an inquiry approach (Figure 2.4).

In this activity, students must figure out how tall a stack of 17 cups is, without being given any more cups. Additionally, they are told that a cabinet in the classroom is 50 centimeters high, and the teacher needs to know how high the stacks should be in order to store the cups. Students will need to decide how to use their tools to determine the height of a stack of 17 cups, and they are encouraged to create a formula for finding the height of n cups instead of using a ruler.

This example gives students a realistic situation that they are encouraged to problematize. They have a set of tools (rulers and four cups) and are asked to create a solution to the problem. No guidance or set of steps is given to them beforehand; they are expected to use their prior experiences and mathematical knowledge to invent a viable solution method. The teacher then encourages them to create a more abstract formula that has quantitative meaning for the students. Later in the unit, they are encouraged to use what the textbook authors term *arrow strings* to solve their invented formulas. When they have found a way to solve the cups tasks, they are invited to share their reasoning in class so that others can critique the viability of their solution *method* as well as the correctness of their *answer.*

Many of the Strands of Proficiency can be inferred here, as students have the potential to develop conceptual understanding that a linear structure comprises an initial unchanging amount (the length of the *hold* [the distance from the bottom of the cup to the bottom of the rim]) and a rate of change (the measure associated with n new cup rims). They are encouraged to use tools to create a model for the problem situation so that the height of the stack can be determined no matter how many cups are stacked (modeling). Unlike in the previous textbook example, students are not given a model to reason with but, instead, are prompted to organize the situation

FIGURE 2.4

An Activity for Writing and Solving Equations from *Mathematics in Context*

STACKING CUPS

Materials:

Each group will need a centimeter ruler and at least four cups of the same size. Plastic cups from sporting events or fast-food restaurants work well.

Measure and record the following:

• The total height of a cup

• The height of the rim

• The height of the hold

(Note: The hold is the distance from the bottom of the cup to the bottom of the rim.)

• Stack two cups. Measure the height of the stack.

• Without measuring, guess the height of a stack of four cups.

• Write down how you made your guess. With a partner, share your guess and the strategy you used.

• Make a stack of four cups and measure it. Was your guess correct?

Source: From *Mathematics in Context* (p. 16), by T. A. Romberg and J. de Lange, 1998, Chicago: Encyclopaedia Britannica. Copyright 1998 by Encyclopaedia Britannica. Reprinted with permission.

themselves and persevere in figuring it out (strategic competence and productive disposition). Because they eventually write and solve formulas, the unit will engage them in developing procedural fluency. Finally, explaining and defending their models and answers is important to the learning of all students in the class (adaptive reasoning).

Another example of strong STEM inquiry comes from a science experiment called Saving Pelicans (Karahan, Guzey, & Moore, 2014). Over six lessons, students develop, test, and revise their models for relocating 600 pelican eggs that were abandoned due to human destruction at a lake area in Minnesota. Students get a "request" from the Minnesota Department of Natural Resources that explains how the eggs were abandoned and asks for students' help in creating and placing enough new nests for the pelicans to be born and survive.

Notice that the students are already being asked to create a solution to a realistic problem, one that has many correct answers, not just one. Through the lessons, students are given a variety of tools and other resources (such as a fictional aerial photo of the pelican eggs) to help them make decisions about how many nests are needed and how to create a new, viable nest for the eggs. For example, students are given an outline of an irregular shape that represents the area of the pelican colonies and are asked to estimate the number of nests in that area (Figure 2.5).

FIGURE 2.5
Aerial View of the Pelican Colony Region

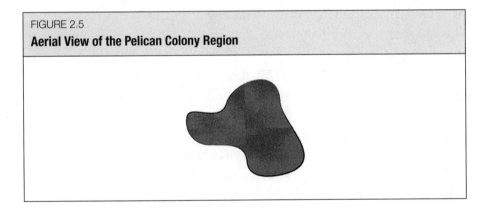

Subsequent activities involve learning about the pelican ecosystem, using the engineering design process to create and test the viability of the nests they manually create (under specific financial constraints), and using geographic information systems technology to make predictions about the best locations to place the nests.

These lessons successfully incorporate student modeling and the design process to test and revise those models. The lessons also incorporate four of the five Strands of Mathematical Proficiency by building conceptual understanding of ecosystems and the effect that humans can have on them, procedural fluency (measuring the area of the pelican colonies), adaptive reasoning (the ability to make a viable argument to the Department of Natural Resources regarding the proposed relocation plan), and strategic competence (creating, testing, and revising a model). Through these relevant and engaging activities, students should develop a positive disposition toward science and modeling, a key idea embodied throughout the Next Generation Science Standards (NGSS Lead States, 2013).

Inquiry Instruction Caution: Teaching for Intellectual Autonomy

Up to this point, we have used the term *inquiry* to describe a method of teaching STEM units and lessons. However, we have found that teachers and researchers use this term in multiple, sometimes conflicting ways. The intent of STEM activities and the instructional supports that accompany them illustrate what we will refer to as *teaching for intellectual autonomy* rather than *inquiry*.

According to Piaget (1948/1973), the major goal of schooling is to promote intellectual autonomy in children, yet the majority of instruction tends to focus on heteronomy, even when using "inquiry" type techniques such as manipulatives, questioning, and even small-group work. Many teachers with whom we work view themselves as inquiry teachers if they use manipulatives, ask a lot of questions, and employ a variety of visual aids in class. While these strategies are crucial to an inquiry approach, many teachers nonetheless use them in a traditional, even heteronomous, way. For example, teachers might introduce the stacking cups activity in Figure 2.4 by telling students to measure the hold of the cup and the rim, and then leading students, through lecture, in how to write a formula. They might pose questions like these:

• How tall is the rim?

- How tall is the hold?
- If we have 17 cups, how many holds is that, and how many rims?
- We can start our equation by writing what for the hold? And x times how many rims?

In other words, the teacher is doing the mathematical thinking for the students. Although he or she is asking the questions and having students conduct the measurements, the students are not creating the solution for themselves. In this way, the teacher is creating a culture of intellectual heteronomy, in that the students come to believe that the more knowledgeable adult is the source of mathematical reasoning and that it is their responsibility to reason in the adult's way, as the authority outside of themselves.

In contrast, the teacher who places a few cups, a ruler, and the problem script in front of a small group of students and asks them to figure out how tall a stack of 17 cups will be is supporting an environment in which students come to believe that *they* are the owners of and authorities over their own learning. Rather than believe that mathematics is created by someone outside of themselves, they become intellectually autonomous—they believe that it is their responsibility to create solutions to problems. If a majority of the teacher's moves involves sharing problems (rather than strategies) with students and providing materials and other pedagogical supports to help them formulate their own solutions, then we say that the teacher is *teaching for intellectual autonomy*.

While the original intent of inquiry was to teach for autonomy, that word has now become synonymous with more structural features of classroom settings, such as providing manipulatives, placing students in small groups, posing problems in a real-world context, and asking questions. We therefore use the term *teaching for autonomy* throughout the remainder of this book to refer to classroom communities that use instructional approaches that encourage students to become the owners of their mathematical thinking and less reliant on the teacher or a textbook to tell them how to solve a problem.

Resources and Organizations

High cognitive demand tasks that incorporate modeling and the teaching for autonomy approach are difficult to find in most commercial textbooks. In this section, we list some resources that we have found to be potentially beneficial for coursework in the STEM areas.

A caveat: given that we live in a digital age, you might think that we would recommend the Internet as the first place to find resources. While we do advocate using the Internet, we also advise caution. People who have not studied instructional design post their activities on the web, sometimes for free and sometimes for a fee. We warn teachers to be very selective with regard to both the sites that resources come from and the instructional activities that are offered. Make certain that the instructional materials you find contain the characteristics we have outlined above. Also, beware of materials that claim to teach a topic (e.g., photosynthesis) in one or two "fun" and "engaging" lessons. Certainly, we want our students to be engaged, and many activities can do that; however, they must be lessons that are situated within a larger unit that incorporates modeling and the inquiry process.

Figure 2.6 lists some examples of websites and materials that we think can be used productively for STEM teaching.

Back to Lesson Imaging

Once you have chosen a unit, how do you unpack the objectives of a given lesson? That is the first question to consider when using the Understanding by Design® framework (Wiggins & McTighe, 2005), whether you are planning just one lesson or a larger unit.

Take another look at the first part of the lesson imaging template (Figure 2.7). We begin with stating the goal or goals for the lesson, which should include either the procedural or the conceptual understanding targeted by the activity. Additionally, whether you are adapting an existing lesson or creating a new one, it needs to be governed by the standards for your state, which is why standards are listed at the top of the template.

FIGURE 2.6
Resources for Teaching STEM

Websites	Textbooks/Journals	Programs
National Council of Teachers of Mathematics: www.nctm.org	*Mathematics in Context* (Encyclopedia Britannica)	Engineering Is Elementary
National Council of Teachers of Mathematics: Illuminations: http://illuminations.nctm.org	*Connected Mathematics Project 3* (Pearson)	LEGO Mindstorms
National Science Teachers Association: www.nsta.org	*Core Plus Mathematics* (McGraw Hill)	Odyssey of the Mind
Achieve the Core: www.achievethecore.org *(classroom resources designed to help educators understand and implement the Common Core and other standards)*	*Interactive Mathematics Program* (It's About Time)	
International Technology and Engineering Educators Association: www.iteea.org	*Project-Based Inquiry Science* (It's About Time)	
NASA: www.nasa.gov	*Science Scope* (journal)	
American Society for Engineering Education: www.asee.org	*Journal of STEM Education*	
International Society for Technology in Education: www.iste.org		
Engineering, Go For It!: www.egfi-k12.org		

To help clarify what to cover, we draw on an example from a 7th grade unit on ratio and rates (Stephan, McManus, Smith, & Dickey, n.d.). The teacher's manual includes lesson images for all the activities within the unit, but begins by defining some of the conceptual goals that serve as targets for the entire unit (listed in the far left column), as seen in an excerpt from a table titled "Big Ideas for the Instructional Sequence" (Figure 2.8).

After the launch page (which we will talk about in the next chapter), the first page that students see asks them to determine if two food bars is enough to feed nine aliens, if one food bar can feed three aliens.

FIGURE 2.7

Part of the Lesson Imaging Template

Science, Technology, Engineering, or Mathematics Goal(s):

State Standard(s):

Cycle 1
Launch (Task presentation)
Exploration (Anticipated student thinking—include class structure [in small groups, with partners, individually] and potential correct and incorrect strategies or solutions)
Whole-Class Discussion (Include tools, symbolizing, technologies, and questions you might pose)

Let's examine just the lesson goal for that page. The teacher page shows anticipated student reasoning (Figure 2.9).

The teacher page also includes notes to the teacher at the bottom:[2]

Big Mathematical Idea(s): The idea of this page is to encourage students to link two composites together.

Rationale: It is an easy page for students, so it should only take a couple of minutes as a beginning page. Make sure to highlight the links that

---

[2]Reading the unit introduction can help answer these critical questions. We encourage readers to download the unit (http://cstem.uncc.edu/sites/cstem.uncc.edu/files/media/Ratio T Manual.pdf) and read pages 5–7 (starting just below the "Big Ideas" table).

	FIGURE 2.8		
	A Snapshot from *Ratio and Rates:* Defining Conceptual Goals		
Big Idea	**Tools/Imagery**	**Possible Topics of Discourse**	**Activity Pages**
Linking composite units	Connecting pictures of aliens to food bars	If the rule is 1 food bar feeds 3 aliens, the rule can't be broken if we add more food bars	Page 1
Iterating linked composites	Informal symbolizing (e.g., tables, two columns of numbers, pictures of aliens and food bars)	How students keep track of two quantities while making them bigger	Pages 2–4

Source: From *Ratio and Rates* (p. 4), by M. Stephan, G. McManus, J. Smith, and A. Dickey, n.d., Oviedo, FL: Lawton Chiles Middle School.

students form between a food bar and a composite of aliens, either verbally or with symbols.

Teachers should consider several questions at this point:

• What does it take to understand ratios and rates at a deep, well-connected level?

• What are the conceptual goals of this first page, and how do they fit within the trajectory of students' learning of ratios?

• The lesson says that it is trying to help students link composite units. What exactly is meant by *linking composite units*?

The teacher's manual notes that some students may solve this problem by circling three aliens and drawing a line from that group of three to one food bar (Figure 2.9). That line connecting two different quantities can be used during discussion to support the idea that those three aliens are linked to one food bar and that link cannot be broken during subsequent problems. So, although the task is very simple, the teacher asks questions to help bring about the lesson objective, drawing students' attention to the fact that they have linked two composite units together.

FIGURE 2.9

A Snapshot from *Ratio and Rates*: Anticipated Student Thinking

Anticipated Student Thinking:

NUMBER ONE

• Some students will say not enough food.
Draw line from one food bar to 3 aliens but the last 3
do not have a food bar.

• Some will circle the entire collection of 3 aliens and
1 food bar

Source: From *Ratio and Rates*, by M. Stephan, G. McManus, J. Smith, and A. Dickey, n.d., Oviedo, FL: Lawton Chiles Middle School.

An Engineering Example

Let's look at another example: a water rockets activity created by NASA, used as the first lesson in an engineering class, in which aerodynamics and Newton's laws of motion are incorporated. Students are given instructions for producing a final product: a rocket created out of a two-liter plastic bottle. Through a series of carefully designed lessons, students engage in cycles of hypothesis, design, data gathering and analysis, and, finally, revision. Their goal is to design a rocket that can fly high, straight, and fast, using the least expensive materials. They are given a variety of tools and materials to work with, and they engage in lessons that encourage them to explore why fins and a nose cone are needed for a rocket and how altering the size and shape of the fins affects the rocket's speed and flight path. They must also consider the cost of the materials and the limitations that cost can impose on design.

There are many ways that students might explore one or more of Newton's laws of motion. For example, young students can flight-test their rockets by filling their bottle rocket one-third full of water and then pumping air into the rocket until the pressure builds, thus releasing the water and sending the rocket shooting off into space. Newton's third law, that every action has an equal and opposite reaction, can be explored by discussing what makes the bottle fly into the air. Older students can explore Newton's second law of motion ($F = ma$) by first shooting off the rocket without water in the bottle, then adding water and seeing the effect of adding mass (water) on the force of the rocket.

Consider the objective for the aerodynamics portion of the instruction:

Lesson objective: Students will understand that the shape of an object can be modified to lessen the air resistance so that the object can flow through the air with more stability.

Teachers might ask themselves: What does it take to understand this idea at a meaningful level? What is the relationship between the shape of an object and the air resistance, and why does this occur? In terms of building a model of a rocket, why does altering the size and shape of the fins and the nose of the rocket increase or decrease its air resistance?

Looking at a representation of a rocket and a plastic soda bottle together with air flow arrows (Figure 2.10) might help us better understand the concept.

FIGURE 2.10
A Rocket and a Water Rocket

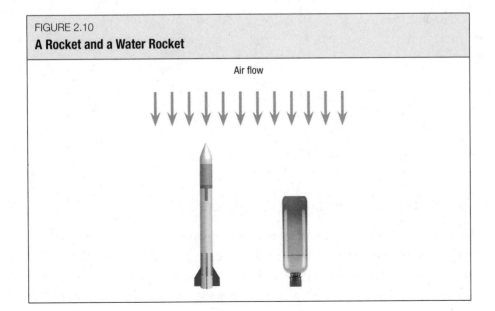

Before looking at Figure 2.11, mentally or physically draw arrows to show how the air would flow around each rocket as it moves upward.

Figure 2.11 makes it clear why a nose cone is used for an actual rocket rather than a flat top. As the rocket travels up through the air with a force, the air reacts against the rocket (Newton's third law). The cone-shaped nose better allows the air molecules to go around the rocket, thus decreasing the resistance. The flat-top rocket does not allow the air molecules to be displaced, which not only slows down the rocket but also creates instability in the flight path.

FIGURE 2.11

Air Flow Around Two Differently Constructed Rockets

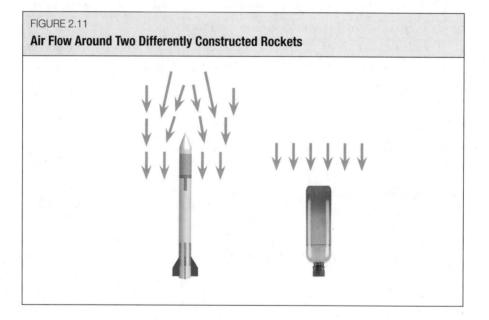

The Importance of Unpacking the Lesson Goals

We remarked in the Introduction that teaching for autonomy often equates to "any exploration goes" in the minds of many skeptics—and in one sense, they are right. Teachers who enact lessons that are not imaged and do not have one or two specific goals in mind will find that the investigations lead everywhere and, consequently, nowhere. If "anything goes," then there is a risk that important standards and objectives are never explored and mastered. Unpacking the lesson goals allows you to anticipate how students might solve the problems or explorations (whether on target with the goal or not) and plan what questions to ask, both to guide students back to the goal and to challenge them by raising the cognitive level of the task.

We agree with Wiggins and McTighe (2005) that understanding the lesson objective is crucial for leading a lesson—and even more so when the lesson is taught with the approach of teaching for autonomy. We return to this topic in Chapter 5, where we explore how knowing the goals of a lesson can help you engineer classroom discussions that both support your objectives and help students develop the five Strands of Proficiency in STEM instruction.

Before Reading Chapter 3...

The next chapter focuses on what it means to image the launch of a task and how to anticipate how students might engage in it. Consider these questions before moving on to Chapter 3:

- What are the characteristics of a good launch for a unit? For a lesson?
- How long should a launch be, and what is the purpose of the launch?
- What does it mean to *model* during a launch, and what is it that is modeled?

3

Imaging the Launch

The launch is most important because this is what connects the students to the lesson and can help them remember what they learn in units down the road. This is the part that sticks in their mind in order to pull out previous information. Especially from a special educator's position, I feel that this is very important because so many times they rely on a connection in order to remember steps or procedures, and this is something that can help them recall information later down the road.

—Erika Allred, middle school special educator, North Carolina

STEM lessons can be introduced in many formats, but the one we highlight here typically follows a three-step process: launch, explore, and summarize (Lappan, Fey, Fitzgerald, Friel, & Phillips, 2013). *Launching* is the term used to indicate the method of introducing a unit or a problem in the classroom. Once the lesson has been launched, the students are given a set time to explore or problem solve, followed by a debriefing summary guided by the teacher but led by students. All three phases of the lesson are discussed in this book, but this chapter focuses on how to image effective launches that engage students in genuine modeling and the inquiry method.

In each phase of lesson imaging, the main goal is to enact the lesson in a way that provides multiple opportunities to learn (Travers, 1993). For us, this means decreasing instances of teacher-driven instruction and encouraging more problem solving from the students. Consider the textbook example from Chapter 2 (Figure 2.3), which asked students to find the solution to the equation $x + 13 = 27$. The launch of this lesson would probably entail the students reading through the page or the teacher explaining the meanings and methods involved in solving the equation. In this way, students are deprived of the opportunity to construct their own method for solving the equation, and, consequently, their autonomy is compromised. In this chapter, we discuss the characteristics of launches that can optimize students' opportunities to construct meaningful and autonomous solution strategies and models.

Characteristics of Effective Launches

There are different characteristics to consider when launching a *unit* versus launching a *problem*. When launching an entire unit that is new to students, some strategies that we have learned from our language arts colleagues can be very effective.

Launching a Unit

Mr. Bell, a 5th grade teacher, decides to use a unit that introduces the notion of variables through the scenario of having students pretend to work in a candy shop and sort candies into rolls and pieces (Underwood & Yackel, 2002). The unit begins with a one-page story about the characters from *The Simpsons*, a popular animated television show (Figure 3.1).

To begin the launch, Mr. Bell informs his students that they are about to begin a unit in which they investigate something called *variables*. He asks students if they have ever heard of the word and, if so, to share some ideas about what it means. Most students say they have never heard of the word, while a couple of students suggest that it means *different*. Mr. Bell thanks his students and acknowledges that although they may not have heard the term before, it does have something to do with *different*. In fact, they are starting a new unit in which certain items take on different values. He then asks if students have ever watched or heard of the television series *The Simpsons*. Students get very excited and start naming some of the characters and favorite episodes. Mr. Bell says that the students are going to read a story that he made up about *The*

Simpsons characters to start the unit. While handing out the first page, he tells students that they are going to work in pairs to *partner-read* this story: one student will read the first paragraph aloud to a partner, and the partner will then reiterate what he or she heard to the reader (Fuchs, Fuchs, & Burish, 2000).

FIGURE 3.1

Introduction to The Candy Shop Instructional Unit

Once upon a time, there was a couple named Homer and Marge Simpson. They had been married for 15 years, when Marge finally said to Homer, "Homey, you HAVE to get a good-paying job soon! How will we ever put Bart through college with the money you make at the nuclear power plant?!" The Simpsons always seemed to be broke, but one day Homer had a brilliant idea. Homer loved to eat Jelly Belly jelly beans, but he thought they were too expensive and he didn't like their chewy consistency (they always got stuck in Bart's braces). He liked hard candies better, but he had never found any that could beat the flavor of a Jelly Belly. Homer asked a local scientist, Mr. Wiz, to help him develop a hard candy with as much flavor as a Jelly Belly. They decided to make their candies disk-shaped like Life Savers, but without the hole.

Uncle Simpson Retires!

As fate would have it, the Simpsons' elderly uncle retired and asked the Simpsons to run his small candy and nut shop. Seeing this as a good sign, the Simpsons took over the business and soon started selling their succulent candies in the candy shop. They renamed their store **The Simpson Sweets Shop**.

The candies were an instant hit! Knowing that people are accustomed to buying candies in rolls—like Life Savers and Rolos—they started to package their candies in rolls as well. But they argued day and night about the number of candies that should be put in a roll.

Drama, Drama!!!

Well, the Simpsons could not come to any agreement about the number of pieces to put in a roll. Marge wanted to put 7 pieces in a roll of orange candy. Homer wanted to put 12 pieces in a roll of cherry candy. But they did agree on one thing: If they put 10 orange candies in one roll, ALL orange candy rolls would contain 10 pieces. And they could put a different amount of candies in a root beer roll—but if they put 17 pieces in one root beer roll, then ALL root beer rolls would have 17 pieces.

Representing Candy in the Candy Shop

Three rolls and 2 extra pieces of candy

Three rolls and a roll missing 2 pieces, or four rolls less 2 pieces

Source: Reprinted with permission from Diana Underwood-Gregg.

After students have partner-read the first paragraph, Mr. Bell stops them
to pose the following questions:

• Who are the characters so far?
• What is the setting?
• What is the plot at this point?

Mr. Bell records their answers on the board. Students state that the
characters are Homer, Marge, Bart, and Mr. Wiz. The setting is unknown
but is possibly the Simpsons' house or a laboratory. The plot so far is that
the Simpsons need more money to send Bart to college, and Homer and
Mr. Wiz collaborate to make a new hard candy.

The partner-read then continues, with the partners switching roles.

After each section ("Uncle Simpson Retires!" and "Drama, Drama!!!"),
Mr. Bell poses some new questions:

• Have any new characters been introduced?
• Has the setting changed?
• Are there any new plot developments?

Again, Mr. Bell writes the students' responses on the board. He makes
sure to elicit what students know about the Simpsons' packing rules:
though each roll can have a different number of candies, *every roll* of a
particular flavor has the same number of pieces. In other words, a grape
candy roll may have 7 pieces per roll and orange may have 10 candies per
roll, but every grape roll has exactly 7 pieces and every orange roll has 10.

Mr. Bell ends the launch by asking students to determine the total
number of candies in each picture:

• If the rolls represent grape candies that are packaged 15 pieces per roll,
how many pieces of grape candy are in the first picture (3 rolls and 2
loose candies)?
• How many grape candies are in the second picture, where one roll is
missing two pieces?

• If the packing rule for orange candies is three candies per roll, how many orange candies are in the first and second pictures?

Students work in their pairs to solve this task and quickly report their answers.

Understanding the meaning of the rolls and pieces representations will be critical for the next investigations in the unit. Students need to understand that the setting for future problems is the Simpsons' candy shop, in which candies are packaged with different packing rules. This is important because students will eventually work with "mystery rolls" whose packing rule is x pieces—if they opened a mystery roll, x pieces would fall out.

Mr. Bell used the language arts strategies of partner-read and listing plot points, characters, and setting in order to engage everyone in the scenario before more extensive problem solving began. He highlighted on the board the most important parts of the story, including the fact that the Simpsons could not decide on one packing rule but that if they decided on, for example, a particular number of grape candies in one roll, *every* roll of grape candies would have that same number of pieces.

Why was this aspect of the plot worth highlighting? Without this understanding, the notion of a variable as representing an unknown number of pieces, but the same number when you define it as *grape*, is critical for understanding that x stands for an unknown packing rule but that each x in the situation represents the same unknown amount.

When teachers lesson image the launch of a unit, one of the most important things to imagine is what aspects of the problem context are critical for success in further problem solving. Partner-reading and whole-class sharing can be extremely helpful, especially for students with mathematics or language disabilities and for English language learners. Teachers might also image that physical props, such as rolls of Smarties, Rolo, or Starburst candies, can be helpful for illustrating that, for example, all Starburst rolls have the same number of candies in the package, but we don't know the packing rule just yet—and a roll of Rolo candies contains a different amount than a Starburst roll, but all Rolo rolls have the same amount in them. At some point, the teacher may show students an "incomplete" roll to illustrate a roll missing two pieces.

Not all units begin with such an extensive story to build the context for problem solving. Some nontraditional textbooks may have a short introduction paragraph and then lead right into solving tasks. In this case, to launch the unit, the teacher might have the students conduct a *chapter tour*, another technique we learned from our language arts colleagues. In this strategy, students independently leaf through the unit, noting highlighted words, bold-faced words, pictures, graphs, and so on. Their goal is to get a sense of what the unit might be about just by looking at the highlights. We have done this with a *Connected Mathematics* unit called "Moving Straight Ahead" (Lappan et al., 2013), which introduces linear equations through the example of walkathons. Students thumbed through the unit and noticed lots of tables and graphs, something about pledge plans, and walking and jumping jack experiments. They noted words they already knew, such as *variable*, *table*, and *graph*, and words they did not, such as *linear equation*. The teacher capitalized on their "noticings" and confirmed that they were indeed going to be thinking about problems involving jumping jack rates, walkathon pledges, and other interesting contexts. The teacher also acknowledged that they would be learning about linear equations and rates of change in the unit.

These two different ways of launching a unit share many commonalities. Most important, the contexts are intriguing to students of that age and motivate them to want to engage in the problem solving they know is coming up. Second, the teacher images some method in which to have students interrogate the text not only by reading but also by analyzing the characters, plots, pictures, words, representations, and so on. Third, neither launch begins the unit with a list of vocabulary words and academic objectives for the students to read. Rather, students are engaged right away in the most important pieces of the context or the unit that will ensure productive problem solving in the long term.

When imaging the launch of a unit, teachers should consider what they will *not* say as much as what they *will* say to introduce students to the long-term or big-picture ideas; revealing too much about how to solve the tasks should not be part of the launch. Teachers should also image what manipulative props or visual aids could be useful for helping students connect with the important aspects of the context, and what reading strategies would enable struggling students to better relate to the text.

Launching a Task

We turn now to imaging the launch of tasks on a daily basis in the classroom. Launching tasks can be very similar to launching a unit, but on a smaller scale. While the launch of a unit can last anywhere from 15 to 25 minutes, launching a task or a set of tasks should take about 3–5 minutes unless there is a more extensive context involved, especially with modeling tasks. According to Jackson, Shahan, Gibbons, and Cobb (2012), there are four important components to consider when imaging the launch of a lesson (see Figure 3.2).

FIGURE 3.2
Key Features of Effective Launches

- Discuss key contextual features
- Discuss key discipline ideas
- Develop a common language to describe key features
- Maintain the cognitive demand

Source: From "Launching Complex Tasks," by K. Jackson, E. Shahan, L. Gibbons, and P. Cobb, 2012, *Mathematics Teaching in the Middle School*, 18(1), pp. 24–29. Copyright 2012 by the National Council of Teachers of Mathematics.

To further explore the ideas, let's examine a civil engineering task that was adapted from *Connected Mathematics 3* (Lappan et al., 2013) to introduce mathematical modeling in linear situations. When imaging the launch of this lesson, a group of teachers envisions projecting a two-minute slide show of different bridges from around the world to hook students into the lesson. What questions might the teacher then pose to get the students thinking about the materials needed to create a bridge, how much weight a bridge can hold, and the various factors that influence the amount of weight a bridge can hold? One teacher suggests showing a video of the Tay Bridge disaster,[1] which occurred in Edinburgh, Scotland, on December 28, 1879—a railcar carrying 75 passengers fell through the collapsed bridge and into the icy waters of the Tay. The teacher imagines that the students will wonder what caused the collapse and, consequently, the death of all 75 passengers. The teacher proposes asking, "What can cause a bridge to collapse?" if the students are not already wondering that aloud. Students might offer such ideas as, "The weight of the train was too much," "The bridge materials were not strong enough," and "Parts of the support were frozen from the winter, and they buckled." After these suggestions, the teacher launches the official task (Figure 3.3).

FIGURE 3.3
Bridge Task Introduction

You work as an engineer for the Eagle Eye Engineering Firm. The Edinburgh City Council has hired your team to build the next bridge over the Tay. Your team decides to build a small-scale model of the bridge and then test various weights on the bridge to determine at what point it will collapse.

What are some factors to consider when building your bridge?

As the next part of the launch, the teachers envision collecting answers to this question, which might include the length of the bridge, height of the bridge, type of material used, and type of design (suspension, beam, arch, etc.). The teachers agree to launch the lesson this way and then introduce the remainder of the task, beginning with a focus on the breaking weight of a bridge and the strength or thickness of the material (Figure 3.4).

---

[1] *The Tay Bridge Disaster* video was created by The Open University in 2009 and is available on YouTube (www.youtube.com/watch?v=YHT_Gz2fJuM).

FIGURE 3.4

Bridge Thickness Experiment

Use the paper bridge strips provided to test the breaking weight of bridges of different thickness (1-, 2-, 3-, 4-, and 5-strip thickness). You will need two books on which to place your bridge. Place each paper bridge, one at a time, one inch from the edges of the books so that the bridge spans the empty space between the books.

Put the paper cup in the middle of the bridge and place pennies in it until the bridge collapses. Record your data for each trial.

Source: Adapted from *Connected Mathematics 3: Thinking with Mathematical Models: Linear and Inverse Variation* (p. 11), by G. Lappan, E. D. Phillips, J. T. Fey, and S. N. Friel, 2013, Boston: Pearson. Copyright 2013 by Pearson.

Students then answer questions like the following about their experiment:

• Is the relationship between bridge thickness and breaking weight linear?

• What would you predict as the breaking weight for a bridge 2.5 layers thick?

• What about 6 layers?

An effective launch of this task should contain all the characteristics outlined by Jackson and colleagues (2012). First, they contend that in an effective launch, the students and teacher discuss important *contextual features*. The teachers envisioned accomplishing this in the launch above by showing pictures of different bridges and talking about the fact that bridges often serve as the connector between cities and sometimes countries. Some bridges can hold different amounts of weight than others, and bridges differ in length, materials, and structure. Most students have crossed a bridge or seen bridges in movies and can readily discuss factors that might contribute to the design and function of a bridge. Students can also watch the slide show again to compare and contrast the different bridges and look for things such as structure, type, length, and height, if need be. Having students make observations about the different purposes, sizes, and structures of bridges makes these design features more realistic, especially to those students who are unfamiliar with bridges and might be lost if these factors are not discussed early in the launch.

A second feature of effective launches is that *key discipline ideas* should be discussed before modeling occurs. Placing students in the role of engineers whose team must create a list of factors that affect the stability of a bridge puts them front and center as designers. They are given the opportunity to conjecture about the various elements of designing a new technology (i.e., a bridge) that can affect its function. Considering the effect that the length and thickness of a bridge have on its stability is what Jackson and colleagues (2012) mean by discussing key discipline (in this case, engineering) ideas that are inherent in the task at hand. Additionally, engaging students in talking through the modeling process is extremely important in this launch. Students should think about how they should put the pennies in the cup (dropping them or placing them gently), how they should judge when the paper bridge has collapsed, and what representation would show the best relationship between the thickness of the bridge and the collapse point.

The third principle that teachers should consider when imaging this task is how to *develop a common language to describe key features*—what terms should be discussed and defined as a group prior to experimentation? For example, terms such as *breaking weight* and *prediction* might be challenging for some students. The teacher and students should develop a common understanding before moving forward with the task.

Finally, it is important that the teachers' image does not include how to solve the task, how to make a graph or table, or how to record and explain the results. For example, if the teacher were to reveal to students that the relationship between breaking weight and thickness is linear, the *cognitive demand* of the task would be compromised. Students' explorations would be reduced to matching the teacher's outcome, rather than exploring what the relationship is and why the data might or might not be perfectly linear. Maintaining the high cognitive demand of the task during launch is extremely important so as not to rob students of the joy and challenge of experimenting and creating explanations for the results they find.

A Mathematics Example of a Lesson Launch

As another example of using Jackson and colleagues' (2012) criteria to image a launch, consider the problem shown in Figure 3.5.

FIGURE 3.5

Using a Mathematics Problem to Launch a Lesson

Ms. Smith's class has decided to participate in the Relay for Life walkathon. Each student must find sponsors to pledge a certain amount of money for each mile the student walks.

The students in Ms. Smith's class are trying to estimate how much money they might be able to raise. What are some different ways to earn money from the sponsors?

Each student found sponsors who are willing to pledge the following amounts:

• Tracy's sponsors will pay $10, regardless of how far she walks.

• Eduardo's sponsors will pay $2 per mile.

• Maria's sponsors will make a $5 donation plus 50¢ per mile.

The class refers to these as *pledge plans*.

Whose pledge plan earns the class the most money? Create some evidence on your paper to defend your conclusion.

Before going any further, create your own image of how you might launch this problem. Keep in mind the key features of effective launches: discussing key contextual features, discussing key discipline ideas, developing a common language to describe key features, and maintaining the cognitive demand.

Your image of the launch for this problem likely differed from what we suggest below, but that is what makes imaging so important. Your students and your classroom context are different from ours; consequently, your image should cater to the students in your context, yet still align to the principles we have listed.

The 7th grade teachers imaging the launch for this lesson began by solving the problem individually to get a sense of what the task required of students. One teacher, Mr. Arb, said that he might have one student read the first paragraph aloud; some of his students have reading difficulties, and hearing the short paragraph read aloud while following along in the text might help. He then envisioned having students reread the paragraph individually and underline words they did not understand or weren't familiar with. Class discussion would begin with a discussion of walkathons: who has participated in one before, what are *sponsors*, how does the program work, what is Relay for Life, and so on. In this way, students who may not be familiar with walkathons could learn from those students who have prior experience.

Mr. Arb's image to this point entails having the students discuss *key contextual features* of the story.

(It is worth noting here that the original *Connected Mathematics 3* problem used the names Ms. Chang, Leanne, Gilberto, and Alana, but the 7th grade teachers changed the names to another teacher in the school and to students who were present in the classroom in order to hook the class and make them more motivated to find the best pledge plan.)

Ms. Nolsheim continued the launch image by suggesting that students read and discuss the pledge plans. What does each pledge plan mean? What does it mean to collect $10, no matter what? What does it mean to have the best pledge plan? These kinds of questions prompt students to discuss the *key mathematical features* and the overall mathematical context of the situation.

Note that the teacher does not plan to explain each pledge plan to the class so they will understand. Rather, the students will attempt to make sense of the pledge plans, and the teacher will listen to ensure that their final interpretation fits with the intention. For example, we have seen teachers launch this task and explain to students that Tracy's table means that she will have $10 every time. An explanation that detailed lessens the cognitive demand associated with making sense of the relationship between money and distance walked in Tracy's pledge plan. Rather, teachers need only recognize that the pledge plan is interesting and leave it to the students to represent the relationship in the table themselves, even if it is incorrect. We discuss in the next chapters why an incorrect solution can actually aid in the development of the mathematical ideas.

In this example, the important feature of *developing a common language* for the key terms in the story might consist of imaging what words the students might need to discuss. During the discussion of the pledge plans,

Ms. Nolsheim wondered if students would bring up the term *fixed amount*. If so, she would capitalize on this contribution because it has important implications for structuring linear equations: the fixed amount is the same as a y-intercept. Another teacher, Ms. Bowman, envisioned needing to discuss words such as *evidence*, *defend*, and *conclusion* for students who still may not be accustomed to the types of explanations and arguments used in the class. Ms. Bowman teaches an inclusion classroom, and she knew that many of her students would likely need to be reminded about what counts as evidence in this unit. Hence, she imagined asking students to brainstorm in small groups various types of evidence they have studied in the chapter so far that might count in this situation. With this approach, she would not tell students to use graphs, tables, or equations, but would provide them with the opportunity to recall those forms of evidence themselves and choose the one that fits their abilities. In a later task, she will ask students to use specific representations, but for now she is interested to see which ones they choose on their own.

Finally, the teachers agreed not to tell students how to make the representations—in particular, they would not "warn" them of silly mistakes to avoid in their representations. For example, Mr. Arb envisioned that some of his students might make a table like the one in Figure 3.6 for Maria's pledge plan.

FIGURE 3.6
A Possible Student Table for Maria's Pledge Plan

Miles	Money
1	$5
2	$5.50
3	$6
4	$6.50

Teachers who expect such mistakes during the launch might warn students that they should start all their tables with zero, not one mile. They could even pass out papers with the outlines of the tables already preprinted with 0, 1, 2, 3, 4, and so on, and then students would just need to fill them out. Or they could hand out the coordinate axes to students with the scales already partitioned and labeled on them and with only the first quadrant shown.

While providing tables and a prepartitioned and numbered graph might save time, it also robs students of the opportunity to analyze the three pledge plans and decide on the appropriate scaling. Providing a set of completely empty tables, graph paper, or a graphing calculator or app on request would be better, and could support students with disabilities by relieving them of the demand to draw the table or graph. However, students would still be responsible for creating the scale and the table entries and deciding what variables to use to label the axes and which column to put them in on the table (left or right). Correct and incorrect decisions by students here would provide an opportunity to discuss dependent and independent variables and where each is placed on the various representations; incorrect numbers in the table could provoke discussions about linearity and nonlinearity and how they show up in the graph. Basically, telling students how to structure their evidence prior to exploration can cheat the students out of the high cognitive level of the task and is contrary to what Jackson and colleagues (2012) have documented as effective launching.

A Science Example of a Lesson Launch

Certain concepts of genetics and natural selection first appear in the 3rd grade science standards (NGSS Lead States, 2013). Standard LS4.B states:

> Sometimes the differences in the characteristics between individuals of the same species provide advantages in surviving, finding mates, and reproducing.

In our next example, a group of 3rd grade teachers meet to create a unit on natural selection. They begin by searching through *Science Scope*, a peer-reviewed journal for middle-level and junior high school science teachers. In one issue, they find a promising set of activities to help students simulate the survival of genetically altered bacteria after being treated to a hand sanitizer that promises eradication of 99.99 percent of the germs it comes in contact with (Welborn, 2013). Together, they brainstorm an effective, stimulating launch using the principles described in the previous sections.

First, they find a 3rd grade–friendly article about the spread of superbugs.[2] They decide to pass out an edited version of the text to pairs of students and have them partner-read a paragraph at a time. After one partner reads the first paragraph and the other summarizes what he or she heard, the teacher will instruct the students to highlight words that are difficult to understand. The teachers expect students to highlight the words *bacterium* and *antibiotics.* In addition, some students might not know the colloquialisms (e.g., a powerful punch) or terms such as *drug-resistant superbugs.* After those words and phrases are identified, the teacher will ask students to explain the meaning and use of each term, only interceding if no student is aware of the meanings. Key language and terms that are critical to understanding the meaning of the text are discussed prior to the activity—*developing a common language*, as described above.

As the class continues to read through the modified passage, key contextual elements should arise in the discussion by students or the teacher. The teachers expect students to note that drugs have been able to kill these disease-causing bacteria in the past but that superbugs have developed a resistance to our medicine. The teacher can provoke curiosity by showing students a picture of a common hand sanitizer that boasts 99.99 percent protection against germs.

The teachers image leading the students into the investigation by pointing out that this hand sanitizer should wipe out virtually all bacteria—so why do infections still occur? The teachers will suggest that the class run a simulation that illustrates why superbugs not only survive the application of hand sanitizer but also multiply and spread (Welborn, 2013). They expect students to be highly engaged in this simulation, which uses mini marshmallows as the bacteria and Skittles as the superbug:

---

[2] "Killer Microbe" is available on the NOVA website (www.pbs.org/wgbh/nova/body/killer-microbe.html).

- On students' "hands" (a paper plate), the teacher will place eight bacteria (marshmallows) and two superbugs (Skittles). The students' task is to determine what it means to kill 99.99 percent of the bacteria on their "hand," as the hand sanitizer claims—how many bacteria and how many superbugs will be killed?
  - Students then receive some "hand sanitizer" (a toothpick).
  - Students will have five seconds to remove the bacteria (both marshmallows and candy) by spearing them with the toothpick and setting them aside, thus simulating the sanitizer wiping out the organisms.
  - Students count how many bacteria, both original and superbug, remain after the first turn. Before the next turn, the teacher doubles the amount of bacteria left on each student's plate.
  - Students do three rounds of the simulation. After the final round, they count how many bacteria—marshmallows and candy—remain on the plate.

The activity concludes by having students identify characteristics of the "superbug" that made it resistant to the "hand sanitizer" (a hard shell) and what made the original "bacteria" susceptible to it (a soft shell), thereby hitting on the standard the teachers identified at the outset of the activity.

In this launch, we can see that the 3rd grade teachers collaborated to engage students in a very realistic context involving the spread of germs. As students read a modified passage, small cycles of discussion occurred to identify critical components of both the context and the science ideas. Furthermore, students identified words they were unfamiliar with, yet whose meaning was essential for understanding the science in the passage and the subsequent simulation. These pedagogical moves illustrate the characteristics of a good science launch as identified by Jackson and colleagues (2012).

Conclusion

In this chapter, we looked at another piece of the lesson image template: the launch. We discussed that effective launches may look different depending on your students but that all good launches contain the same four elements: discussing the key context features, discussing key discipline ideas, developing a common language to describe key features, and maintaining the high cognitive demand of a task by not giving away solution strategies. Launches like this maintain the integrity of the problem-solving intent while enabling students to participate in meaningful ways. Powerful launches hook students (Wiggins & McTighe, 2005), provoke them to solve realistic problems that hold their interest and are worth solving, and do not steal the joy of creation, observation, and analysis from the students.

Before Reading Chapter 4...

In the next chapter, we look at one of the most difficult parts of lesson imaging: anticipating students' solutions and solution strategies for challenging problems. Consider these questions before moving on:

• What resources can a teacher draw on to better anticipate students' reasoning?

• Do you use research to guide your instruction? If so, how?

• When you think about student solutions to tasks, do you consider only misconceptions and mistakes? How do you use students' mistakes to further your instruction?

4

Imaging Student Reasoning

The main challenge associated with anticipating students' thinking is being able to switch from a teacher's perspective of the mathematics to the students' perspectives. When I first started teaching this way, I could only anticipate one or two different ways students might think, but with more experience and imaging with colleagues who think differently, I am getting better at it. Anticipating students' thinking helps teachers structure the lesson based upon student responses as opposed to presenting the lesson the way the teacher understands it.

—Ashley Dickey, middle school mathematics teacher, Florida

All good teachers, no matter what instructional techniques they use, try to anticipate their students' misconceptions. With direct instruction, the teacher might structure the lecture in such a way as to prevent mistakes; however, an inquiry teacher uses mistakes as opportunities for student learning and incorporates them as a critical part of the lesson. Rather than attempt to prevent misunderstanding, the inquiry teacher images what might occur naturally in the problem solving and how the teacher can capitalize on it as a learning opportunity for all students. Thus, anticipating how students will reason about a problem includes imaging how both the conceptions and the misconceptions might become public during a whole-class discussion.

In the previous chapter, we discussed several characteristics of a successful launch, one of which is that students should be motivated to engage in tasks that allow for a variety of solution methods that are not generated by the teacher. Anticipating students' solution processes and reasoning is the next portion of the lesson image and can be seen in Figure 4.1.

FIGURE 4.1
Partial Template for Imaging the Exploration Period

Launch (Task presentation)

Exploration (Anticipated student thinking—include class structure [in small groups, with partners, individually] and potential correct and incorrect strategies or solutions)

Several questions emerge about next steps after teachers image the launch:

- What is the teacher's role *after* the launch, when students are exploring solutions to the problem?
- What if students solve the problem in a way we do not understand?
- What if a student can find no way to do the problem?
- What do we do when we see a mistake or a path that is not productive?
- Can we help students if they are stuck?
- How much can (or should) we tell them?

Monitoring students' explorations in ways that preserve their autonomous work is difficult, yet it is critical for the next phase of the lesson. In this chapter, we look at a 5th grade classroom learning about volume for the first time, and we discuss the resources their teacher drew on to make anticipating students' thinking easier and more powerful.

Imaging a Lesson on Volume

The partial lesson image shown in Figure 4.2 was created for a 5th grade classroom of students who had not been introduced to volume prior to this unit. The teachers chose a series of tasks that they developed by reading an article on cognition-based assessment and elementary school students' understanding of volume (Battista, 2004) and using a Java minitool they found online.[1] Their goals for the first lesson are listed in the lesson image template (Figure 4.2).

The teachers formed these goals by referring to the learning trajectory discussed in the Battista article. Battista argues that students develop an understanding of volume by forming mental models—in this case, spatial images—that coordinate units into a three-dimensional (3-D) array. In other words, students can eventually interpret a shape like the one in Figure 4.2 as consisting of a two-dimensional (2-D) array of squares (a 2 × 3 section on the bottom) that enumerates two rows of three squares forming the base of cubes, which are structured in a 3-D array. Through a process of abstractions, eventually students can interpret a 3-D shape (represented as a 2-D shape on the board or computer) as a set of composite units that are coordinated together. For example, in Figure 4.2, a student may see the front "slice" of six cubes (one composite of six) and iterate that slice three times back for a total of 18 cubes.

After reading this and other articles about volume (e.g., Battista & Clements, 1996), the teachers decided to begin the unit with a realistic situation in which students worked for a fictitious candy factory and were in charge of filling boxes with cube-shaped candies. The first few tasks involved projecting the inside of a box of candies on the board and asking the candy factory staff how many candies the box contained. Students were allowed to come close to the board if they needed a better view of the shape. They were also permitted to create the shape with wooden one-inch cubes at their

---

[1]"Cube houses" is available on the Freudenthal Institute's Games for Mathematics Education website (www.fisme.science.uu.nl/publicaties/subsets/rekenweb_en/).

FIGURE 4.2

Partial Lesson Image and First Candy Box Shown to Students

Science, Technology, Engineering, or Mathematics Goal(s):

Introduce students to the notion of packaging cube-shaped candies in a box and determining the number of cubes needed to fill it. We will work on counting how many cubes fill the box with no overlaps and gaps.

Simultaneously, students' spatial sense will be enhanced as they work with cube candies that are shown via a computer program; they will have to count cubes (3-D objects) that are represented two-dimensionally on a screen.

The notion of forming "slices" that can be iterated along a dimension is the end goal of the next few lessons, which forms the foundation for interpreting the volume of a prism as the area of a base (i.e., slice) times the height (iterations along a dimension).

State Standard(s): CCSS.Math.Content.5.MD.C.3: Recognize volume as an attribute of solid figures, and understand concepts of volume measurement.

• A cube with side length of one unit, called a "unit cube," is said to have "one cubic unit" of volume, and can be used to measure volume.

• A solid figure that can be packed without gaps or overlaps using n unit cubes is said to have a volume of n cubic units.

Launch (Task presentation)

Students will be told that they are working for the Chocolicious Candy Company. Their job is to fill boxes with cube-shaped candies and to determine how many candies will fill the entire box.

Give students the handout to partner-read. Make sure that students circle any words in the story they do not understand, and discuss those words as a class.

An important point to include in the launch is that the workers pack the boxes with no missing candies (no gaps).

Show students the bag of "candies" (wooden 1" × 1" × 1" cubes) and 1" grid paper at their tables, and let them know they can use the cubes to solve some of the problems.

Pose the first problem by showing the first candy box on the board (using www.fisme.science.uu.nl/publicaties/subsets/rekenweb_en/) and asking students to build it with their cubes.

Exploration (Anticipated student thinking—include class structure [in small groups, with partners, individually] and potential correct and incorrect strategies or solutions)

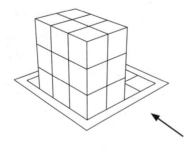

table. The teacher would first present candy boxes that were small enough that children could count by ones and create the shape with cubes fairly quickly if need be. Students were also given one-inch grid paper on which they could draw the base of the box if they chose to. Students who did not need to create the shape with cubes would be encouraged to find easy ways to compute the number of candies from the picture. The candy boxes would eventually get larger, and students would no longer have enough cubes at their table to build the entire shape, but they would be able to build a base or front layer to be iterated.

The teachers decided that students would work in pairs to solve the task. The teacher would walk around the room, monitoring the students' reasoning, and recording the diversity of ways in which the students were reasoning on a tablet.

At this point in the imaging process, the teachers discussed how they thought students might reason as they determined the number of candies contained in the first candy box. Each teacher attempted to solve the problem in as many different ways as possible, with and without the wooden cubes and grid paper.

Before reading the strategies they anticipated, try making your own predictions and then compare yours with what they hypothesized. We always recommend solving the task both with and without the physical materials in order to get a more genuine sense of the ways that students might reason. The teachers hypothesized five different ways of reasoning (Figure 4.3).

This list of hypothetical strategies better equips the teacher to lead the follow-up whole-class discussion, as there is less chance that the teacher will be surprised by students who thought about the problem differently than the teacher did. The teacher is also more prepared with instructional supports at the ready (e.g., the grid paper and cubes) for helping students structure their thinking. Anticipations like these are critical for the success of the lesson for a number of reasons:

• Hypothesizing how students might participate in the task in diverse ways can aid in the differentiation process. For example, some students, particularly those with spatial disabilities, may not have formed the cognitive structures for seeing "slices" and may need to create the physical shape for many problems. Knowing this, teachers can provide the grid paper for students as a structural and visual support for those layers, especially the bottom layer of six. Drawing an outline of the six bottom cubes or coloring in those bottom six squares can help all students visualize a 2-D array of a slice (or, as Battista would call it, a *composite of six*).

FIGURE 4.3
Anticipated Student Strategies for Solving the First Problem

Exploration (Anticipated student thinking—include class structure [in small groups, with partners, individually] and potential correct and incorrect strategies or solutions)

Strategy A: Students will make the correct shape with cubes and count them by ones to get 18 cubes.

Strategy B: Students will make the correct shape with cubes and miscount the cubes by ones because they double-count some of them by pointing to different surfaces of the same cube.

Strategy C: Students will build the bottom layer of six first (using six squares on the grid paper, or not) and notice that they have three layers of six. They will multiply 3 cubes by 6 cubes to get 18 cubes.

Strategy D: Students will build the front face [indicated by the arrow in Figure 4.2] of six (using two squares on the grid paper, or not) and notice that they have three layers of six. They will multiply 3 cubes by 6 cubes to get 18 cubes.

Strategy E: Students will build the side view of nine cubes (using three squares on the grid paper, or not) and then recognize that there are two layers of nine. They will multiply 2 cubes by 9 cubes to get 18 cubes.

NOTE: If students have difficulty creating a physical shape from the image on the board, the teacher can click on the arrow in the picture to rotate the candy box, which gives students a more visual sense of the entire box (the back side, underneath, etc.).

• Having an idea of the variety of ways that students might participate, and making a record of which children use which strategy, can be helpful for creating a portfolio of each student's mathematical growth over time. For instance, during explore time, as students consider ways to solve the problem, the teacher can record on an iPad or other device the particular strategies the teacher observes students using. Over time, the teacher can pull up a long-term record of any student to show parents during a conference and to help students evaluate their own progress. This also helps teachers provide more targeted individual attention to struggling students.

• Creating a list of possible solution strategies, both correct and incorrect, is incredibly important for leading the whole-class session that occurs in the next phase of the lesson. Rather than let the summarize time consist of students sharing in a random show-and-tell fashion, possibly losing other students' interest, the teacher can better structure a discussion and/or debate that engages students and helps bring the mathematical intent of the daily lesson to the fore much more powerfully. (This benefit is the subject of the next chapter.)

A Few Words About Monitoring

As teachers monitor students' explorations, it is important that they see themselves in the role of *data collectors*, rather than *problem-solving experts*. The exploration should be a time when students can genuinely attempt new solution processes without fear of being penalized or corrected. If the teacher approaches a pair of students who have miscounted the blocks, the teacher should ask the students what Mehan (1979) calls an *answer-unknown question*—a question for which the teacher does not know the answer (for example, "How did you get 20 for your answer?"). These questions tend to be genuine inquiries into students' thinking in order to understand the reasoning that led to their solution. In contrast, many teachers ask more *answer-known* questions, such as, "What is the correct answer to this problem?" Not only does the teacher know the answer to the question, but the children know that the teacher has the correct answer in mind and that they are supposed to provide it.

We have found that asking answer-unknown questions during explore time alleviates the threat that students associate with attempting new strategies, inventing creative solution processes, and making mistakes. Students infer that teachers are interested in knowing their thinking, rather than evaluating them. Answer-unknown questioning also facilitates students' growth of intellectual autonomy, as they learn that their solutions are correct only if they are able to justify their solutions, not just repeat someone else's steps.

These types of questions also allow the teacher to diagnose the students' reasoning before following with further instructional support.

We have personally been guilty of asking students what the correct answer is, and then, when we get an incorrect answer, picking up the wooden cubes ourselves and building with or counting the cubes to "show" students how to count. If you find yourself with your hands on the materials or a student's pencil in your hand, it is very possible that you are taking over the problem solving rather than assessing students' thinking. Most importantly, *it is OK to let students finish problem solving with a mistake during explore time*, as long as those mistakes and the invalid reasoning are discussed during the whole-class discussion so that students have an opportunity to know why their solution is being challenged and hear other interpretations that are valid.

If teachers view themselves as data gatherers during explore time, then the follow-up discussion with the whole class can be much richer and deeper mathematically. However, if the teacher sees a student struggling or notices that someone has miscounted the cubes and stops to "fix" it, the whole-class discussion can suffer as a consequence. Imagine a whole-class discussion in which everyone had the exact same answer and used one of only two different strategies. There wouldn't be much to talk about because the solutions and strategies are so homogeneous. Mistakes often provide fodder for important mathematical discussions, and eliminating them during explore time steals the opportunity for others, even those with correct answers, to learn from them.

For example, hearing the reasoning of students who miscounted the cubes and got 20 can be beneficial to the whole class. Having two different answers on the board—say, 20 and 18—provokes interest in all students, because they want to know which answer is correct. Students are more likely to listen to justifications if they have to decide which one is correct by the end of the discussion. They are also more likely to share their reasoning, sometimes very passionately, when they are attempting to prove that their answer is the right one.

Additionally, important mathematical ideas related to structuring the spatial array can come to the fore when students discuss their strategy for "finding" 20. As other students try to make sense of this solution process, they will probably argue that the students who got 20 must have counted a couple of cubes twice. They might even come up to the board to show that the square on the front of the shape and the square on top of the shape each represent the same cube (Figure 4.4).

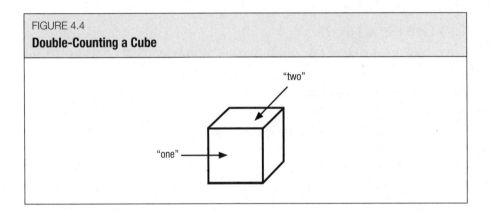

FIGURE 4.4
Double-Counting a Cube

Such a way of counting may not have occurred to others who got 18 as the answer, bringing to their attention that there is a way to structure and enumerate the cubes in a 3-D shape.

Resources and Practices for Improving Anticipations

According to several researchers (see, e.g., Cobb, Yackel, & Wood, 1989; Inoue & Buczynski, 2011; Stephan, Underwood-Gregg, & Yackel, 2014), teachers who are new to teaching for autonomy can be extremely successful in creating safe, risk-free environments where students listen to one another and explain their reasoning. However, anticipating students' thinking *before* you have taught a particular lesson is one of the most difficult parts of teaching for autonomy. This is particularly true for preservice teachers who haven't yet worked with students and are not starting with the experience that most classroom teachers have. Nonetheless, even experienced classroom teachers can have difficulty anticipating, because it is difficult to "get into the heads" of a variety of students. In our experience, teachers who are new to this practice are able to anticipate, at most, two or three solution strategies for each lesson. What practices and resources can increase the success of predicting how students may reason?

The likelihood of anticipating most student reasoning increases when teachers do the following:

- Lesson image together
- Read research that explains the cognitive development on that particular subject
- Work through the tasks themselves, with the manipulatives if appropriate
- Pre-assess students' knowledge with cognitive interviews

Lesson Image with Colleagues

While this may seem very obvious, it is rarely done in practice. As we will discuss in Chapter 7, rarely do school schedules allow common planning time for teachers to lesson image with peers. Even when there is common planning time, not all teachers understand how to lesson image and often talk instead about other issues, such as the behavior of shared students. In order to make lesson imaging more productive, we suggest that teachers who know how to image, or are willing to learn, form a small professional learning group. In this way, teachers can capitalize on the diversity of thinking of other adults.

Read the Relevant Research

Consider the example of the elementary school team who lesson imaged together for a sequence of lessons on volume. Crucial to their anticipations was *reading research* on how students develop their understanding of volume. The articles they read attempted to articulate what might be called a *learning trajectory* for volume, or a set of cognitive constructions that are made as students learn for understanding. We recommend searching for articles that use the term *learning trajectory* or focus on students' cognitive development of a particular STEM area. Figure 4.5 lists a few examples of journals, books, and articles that can be used to help with anticipations.

FIGURE 4.5

Helpful Publications for Anticipating Student Thinking

Journals	Books and Articles	Articles
Journal for Research in Mathematics Education	Glynn, S., & Duit, R. (Eds.). (1995). *Learning science in the schools: Research reforming practice.* New York: Routledge.	Kilpatrick, J., Martin, G., & Schifter, D. (2003). *A research companion to Principles and Standards for School Mathematics.* Reston, VA: National Council of Teachers of Mathematics.
Journal of Research in Science Teaching	Carmichael, P., Driver, R., Holding, B., Phillips, I., Twigger, D., & Watts, M. (1990). *Research on students' conceptions in science: A bibliography.* Leeds, UK: Children's Learning in Science Research Group, CSSME, University of Leeds.	Blanton, M., Brizuela, B., Gardiner, A., Sawrey, K., & Newman-Owens, A. (2015). A learning trajectory in 6-year-olds' thinking about generalizing functional relationships. *Journal for Research in Mathematics Education, 46*(5), 511–558.
Journal of Engineering Education	Black, P., & Lucas, A. (1993). *Children's informal ideas in science.* London: Routledge.	Clements, D. H., Wilson, D. C., & Sarama, J. (2004). Young children's composition of geometric figures: A learning trajectory. *Mathematical Thinking and Learning, 6*(2), 163–184.
Journal of Technology Education	Abraham, M., Williamson, V., & Westbrook, S. (1994). A cross-age study of the understanding of five chemistry concepts. *Journal of Research in Science Teaching, 31*(2), 147–165.	Clements, D. H., & Burns, B. A. (2000). Students' development of strategies for turn and angle measure. *Educational Studies in Mathematics, 41*(1), 31–45.
Journal of STEM Education	Aranudin, M., & Mintzes, J. (1985). Students' alternative conceptions of the human circulatory system: A cross-age study. *Science Education, 19*(5), 721–733.	Stephan, M., & Akyuz, D. (2012). A proposed instructional theory for integer addition and subtraction. *Journal for Research in Mathematics Education, 43*(4), 428–464.
Journal of Educational Technology and Society	Bar, V., & Travis, A. S. (1991). Children's views concerning phase changes. *Journal of Research in Science Teaching, 28*(4), 363–382.	Westbrook, S. L., & Marek, E. A. (1991). A cross-age study of student understanding of the concept of diffusion. *Journal of Research in Science Teaching, 28*(8), 649–660.

FIGURE 4.5—(continued)

Helpful Publications for Anticipating Student Thinking

Journals	Books and Articles	Articles
NCTM Journals • *Mathematics Teacher* • *Teaching Children Mathematics* • *Mathematics Teaching in the Middle School*	Beveridge, M. (1985). The development of young children's understanding of the process of evaporation. *British Journal of Educational Psychology, 55,* 84–90.	Stavy, R., Eisen, Y., & Yaakobi, D. (1987). How students aged 13–15 understand photosynthesis. *International Journal of Science Education, 9*(1), 105–115.
Prism (American Society for Engineering Education)	Novick, S., & Nussbaum, J. (1981). Pupils' understanding of the particulate nature of matter: A cross-age study. *Science Education, 65*(2), 187–196.	Séré, M. (1982). A study of some frameworks in the field of mechanics used by children (aged 11 to 13) in the interpretation of air pressure. *European Journal of Science Education, 4*(3), 299–309.
Journal of Technology and Teacher Education (Society for Information Technology and Teacher Education)	Nussbaum, J. (1979). Children's conceptions of the earth as a cosmic body: A cross-age study. *Science Education, 63*(1), 83–93.	Osborne, R. J., & Cosgrove, M. M. (1983). Children's conceptions of the changes of state of water. *Journal of Research in Science Teaching, 20,* 825–838.

Try the Task Ahead of Time

Again, this suggestion may sound obvious, but sometimes teachers take it for granted that they know how to solve a problem. We have found that the number of diverse anticipations rises when we explore the problem ourselves with an eye toward thinking of different ways to engage in the task.

We have also come upon instances where certain manipulatives are needed for the task, and if we had not solved it with the actual physical materials, our lessons would not have succeeded. For example, recall the task in Chapter 2 (Figure 2.4) that involved eventually writing an equation to see how tall a stack of cups would be depending on the number of cups used. Students were to measure the rim and the hold of the cup and then stack them to see how tall the stack would be.

Initially, we were just going to anticipate how students might solve the task without conducting the physical measuring ourselves. However, one of the mathematics facilitators pushed us to do so, and we found very quickly that if the students used centimeters rather than inches, the task would go in a direction that was counterproductive to the lesson goal (i.e., the discussion would involve the precision of measurement, which is an important goal, but not for this lesson). In fact, when using rulers, the rim of the cup was approximately 1.5 inches, which led us to anticipate different solution strategies than if the students had had to decide on a number between 3.7 and 3.9 centimeters. We would never have figured that out if we had not worked through the task ourselves, using rulers.

Pre-Assess Students' Knowledge with Cognitive Interviews

Conducting *pre-unit cognitive interviews* can help you learn how students might approach a variety of tasks, which allows you to better anticipate students' reasoning. We have written about this process elsewhere (Stephan & McManus, 2013; Stephan, McManus, & Dehlinger, 2014); here, we offer an example of how one learning community approached this task.

The 6th grade mathematics teachers at a Florida middle school, who were to begin teaching fraction multiplication in about four weeks, had a question: what does it mean to understand fraction operations at a conceptual level so that the rules for multiplying and dividing make sense and are not simply procedures to memorize without necessarily being understood?

The teachers knew that pre-assessing students to see what they already understood about the topic could help them decide where to start in the textbook. However, their mathematics coaches suggested going beyond traditional pretests, which typically consist of questions that elicit procedures and right or wrong answers, with little insight into students' thinking. The coaches recommended that prior to teaching, the faculty conduct *cognitive interviews* to understand how students currently structure fractions and make sense of multiplying and dividing fractions. The coaches also did a literature search for articles about how students understand fraction operations and found several that were very helpful (Figure 4.6).

From their reading of the research, the math coaches put together a set of interview questions (Figure 4.7).

FIGURE 4.6
Research Publications on Fraction Operations
Damarin, S. (1976, April). *A proposed model for teaching and learning common fractions, and the operations of multiplication and division of fractions.* Paper prepared for the Georgia Center for the Study of Learning and Teaching Mathematics Workshop on Models for Teaching and Learning Mathematics, Atlanta, GA.
Empson, S., Junk, D., Dominguez, H., & Turner, E. (2005). Fractions as the coordination of multiplicatively related quantities: A cross-sectional study of children's thinking. *Educational Studies in Mathematics, 63,* 1–28.
Hardiman, P., & Mestre, J. (1987). *Understanding multiplicative contexts involving fractions.* (ERIC Report ED290628.) Washington, DC: Department of Education.
Mack, N. (2001). Building on informal knowledge through instruction in a complex content domain: Partitioning, unit, and understanding multiplication of fractions. *Journal for Research in Mathematics Education, 32*(3), 267–295.
Taber, S. (1999). Understanding multiplication with fractions: An analysis of problem features and student strategies. *Focus on Learning Problems in Mathematics, 21,* 1–27.

As the teachers finished teaching their current unit, the mathematics coaches interviewed a diverse set of students in order to get a general sense of how the 6th graders were reasoning about multiplication and division. For example, of the students that were pre-interviewed, some had mathematics disabilities, some had not been diagnosed with a learning disability but were failing the state mathematics test at a low level, some students were performing at a proficient level but on the bubble, some had a strong proficiency score, and the others were either gifted or in honors mathematics classes. Such academic diversity was important to all teachers so that a wide range of student reasoning could emerge from the interviews and thus a richer collection of student reasoning could be collected. The teachers were encouraged to attend any interviews that were conducted during their planning period. The coaches organized their interview results and shared them with the 6th grade team during a half-day professional development meeting arranged by the principal.

When the coaches analyzed students' strategies during problem solving, they found five common themes:

• Most students had difficulty renaming the fraction in terms of a different unit. For example, when solving the first problem ("Name the shaded

FIGURE 4.7

Interview Questions and Rationales

#	Question	Form	Rationale
1	Name the shaded region. Can you call it 1/2 of something? Can you call it 1/3 of something?		Can students name the size of the shaded area and reason in flexible ways, using different units as the referent?
2	Willy Wonka is giving free candy to the first five children who visit his candy factory on Tuesday. If he gives 2/3 of a chocolate bar to each child, how many candy bars will he give away?	$n \times b/c$	How do students solve problems of the form *whole number times a fraction?* Do students take 2/3 as an object to be iterated multiple times?
3	You have 1/2 of a chocolate chip cookie. You give Tonya 1/4 of the cookie you have. How much of the whole chocolate chip cookie did you give your friend?	$1/a \times 1/b$	How do students solve problems of the form *unit fraction times unit fraction?* Can students take a part of a part and rename it in terms of the whole?
4	Pam went to a birthday party and brought home some leftovers: 3/4 of a cake. Matthew ate 2/3 of the leftovers. How much of the whole cake did Matthew eat?	$a/b \times b/c$	Can students conceptualize fractional amounts as embedded within a unit and not partition the unit any further?
5	George has 2/3 of a candy bar. He gives 3/4 of his amount to his daughter. How much of a whole candy bar did George give to his daughter?	$a/nb \times b/c$	Can students reconceptualize a unit by repartitioning the partition?
6	Donna has 9/10 of a candy bar. She gives 2/3 of her amount to her son. How much of a whole candy bar did Donna give to her son?	$a/b \times nb/c$	Can students reconceptualize a unit by regrouping pieces of the original partition?
7	Michelle has 7/8 of a candy bar. She gives 3/4 of her amount to her cousin. How much of a whole candy bar did Michelle give to her cousin?	$a/b \times c/d$ where b and c are relatively prime	Can students reconceptualize a unit by repartitioning the unit and renaming the resulting pieces?
8	Show students a paper with the four number sentences below, and ask them to write a story problem for each one. $4/5 - 1/8 = ?$ $4/5 \times 1/8 = ?$ $4/5 + 1/8 = ?$ $4/5 \div 1/8 = ?$		How do students interpret the abstract symbols associated with fraction operations?

region"), most students argued that it could not be called 1/2 of anything because they had difficulty using anything other than a whole circle as the referent unit. The teachers knew that this conceptual idea is crucial to understanding multiplication and division of fractions: 1/2 of 1/2 is equivalent to 1/4, and the shaded region in the circle is 1/2 of a *half* circle and 1/4 of a *whole* circle. Hence, understanding multiplication (and division) of fractions involves renaming an amount with different units.

• Analyzing students' solutions to the fourth problem led teachers to a startling revelation about the form of the fractions used in the problems: when the *denominator* of the first fraction and the *numerator* of the second fraction are the same (as in 2/3 of 3/4), there is potential to be correct by accident when reasoning with a continuous unit, like the area of rectangles (naming pieces, not fractions).

This "aha" moment occurred while listening to Jill reason about her solution. Jill drew four pieces inside a rectangular cake, due to the prompt that Pam had brought home 3/4 of a cake (Figure 4.8a).

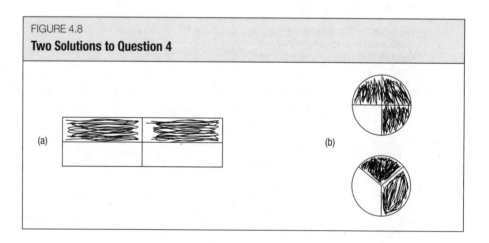

FIGURE 4.8
Two Solutions to Question 4

(a)

(b)

Jill did not attempt to shade in three of those pieces (to represent 3/4); instead, she just colored in two of them, because the prompt said that 2/3 was eaten. Hence, the teachers summarized that Jill was reasoning about whole pieces and not reasoning about thirds of fourth-sized pieces. Another way to state it is that Jill was not reasoning with fractions but, rather, whole-number pieces.

This surprise finding led the 6th grade teachers to conclude that they should provide multiplication problems that used all the forms listed in Figure 4.7. They would also be more conscious of the form of the fraction rather than choose numbers randomly or use examples from the book, in order to ensure that students are working with different forms.

• Students misinterpreted a division situation as a subtraction situation, as revealed by the story problems they wrote for question 8. For example, for $4/5 \div 1/8$, many students wrote something like this: Craig has 4/5 of a brownie and shared 1/8 of a brownie. How much does he have?

• Students had difficulty finding and understanding part of a part. This led the teachers to understand that taking parts of parts and subtraction would both need to be compared during instruction.

• Students had difficulty partitioning twice (double-partitioning) on the same diagram. This is related to their difficulty with understanding a part of a part, as shown in Chanlar's strategy for problem 4 (Figure 4.8b). Chanlar drew a round cake, partitioned it into thirds, and colored in 2/3 of it. Above it, she drew another cake of the same size, and colored in 3/4 of it. Chanlar created two different drawings with each partitioned into the size she needed, rather than one drawing that was partitioned twice—she seemed unable to coordinate the two different partitions together. The teachers gleaned from this that students would have difficulty partitioning a partition (i.e., taking a part of a part).

After analyzing these themes, the 6th grade teachers felt more prepared to anticipate student reasoning, as they now had a palette of strategies that students might use to solve problems in the upcoming unit. The teachers were also better able to approximate a reasonable starting place for instruction and to choose targeted problems or tasks. More specifically, the fact that many students could not rename a region using a different unit told teachers that they needed to revisit the meaning of a fraction. Therefore, when thinking about their first lesson in the multiplication-division unit, they created the lesson image shown in Figure 4.9.

FIGURE 4.9

Anticipated Thinking for the First Fraction Multiplication Lesson

Science, Technology, Engineering, or Mathematics Goal(s):

Students will understand that a shaded region can have different fractional names depending on the unit of reference.

State Standard(s): CCSS.Math.Content.5.NF.B.4: Apply and extend previous understandings of multiplication to multiply a fraction or whole number by a fraction.

Launch (Task presentation)

Students will be shown a picture of a shape with a smaller region shaded in (all shapes similar to question 1 in Figure 4.7). Students will be asked, "What fraction would you name this shaded region?"

NOTE: Do *not* ask students, "What fraction of a circle is the shaded region?" as that would only allow for one referent unit.

Exploration (Anticipated student thinking—include class structure [in small groups, with partners, individually] and potential correct and incorrect strategies or solutions)

The anticipated student reasoning below is based on solutions to the first problem in the pre-interview (Figure 4.7).

Solution 1: Students might call it 1/4.

Solution 2: Students might call it 1/3.

Solution 3: Students might call it 2/8, if they imagine cutting the circle into more pieces.

Solution 4: Students might give other equivalent names by adding more lines.

Solution 5: Students might call it 1/2.

Solution 6: Students might call it 1.

This activity will be whole-class discussion only, with students individually writing down their solutions on paper and then offering their answer and rationale when called on.

NOTE: Be sure to get 1/3 into the discussion, so that equal-sized pieces are talked about. Also, focus on 1/2 and 1, so that changing the referent units is discussed.

Further Practice Anticipating Student Thinking

As another example of anticipating student thinking, consider an engineering lesson that involves building towers out of pipe cleaners. There are three learning objectives that students should understand after doing this activity:

- The concept of limited resources and constraints
- The importance of teamwork and communication
- The importance of planning a project

Before looking at our version, find the lesson online[2] and attempt the task by yourself or with peers. Then, fill out a lesson image template all the way through "Exploration". Our lesson image is shown in Figure 4.10.

Were you able to anticipate how students might solve this problem? Did you try to anticipate students' reasoning alone or with peers? Did you solve the problem mentally or with actual pipe cleaners?

We contend that engaging in the practices we outlined above (lesson imaging with peers, reading research on learning trajectories and cognition, trying the task yourself, and conducting interviews to elicit students' pre-unit informal strategies) can improve your ability to predict how students will participate in problem solving successfully.

For further practice, we offer another example—in this case, a science lesson on the origins of different skin colors and pigmentations (Prud'homme-Généreux, 2011). The launch of the unit involves "hooking" students by asking them about their prior experiences with skin cancer: "Do you know anyone who has been treated for skin cancer? What was that person's experience like?"

The teacher then introduces the unit by having students read a short scenario about a dermatologist and his daughter. To generate the semantic grounding for the investigation, students are asked to work in small groups to consider the causes of skin cancer. In this unit launch, students are simply relying on prior knowledge or experiences to hypothesize the causes of skin cancer and why it appears most often in Caucasians.

Read through the launch of the unit (Figure 4.11), and list all the responses you can think of that your students might give for the questions posed.

For the purpose of the unit launch, it is not important that *all* the anticipated responses come out in the small-group or whole-class discussions;

---

[2]Pipe-Cleaner Towers is available on the eGFI website (http://teachers.egfi-k12.org/pipe-cleaner-towers/).

however, responses that involve sunshine and melanin should be highlighted because they relate to the goals of the unit and future lessons.

FIGURE 4.10

Partial Lesson Image for Pipe-Cleaner Towers, an Engineering Lesson

Science, Technology, Engineering, or Mathematics Goal(s):

Students will understand:
- The concept of limited resources and constraints
- The importance of teamwork and communication
- The importance of planning a project

State Standard(s):
Common Core Mathematical Practice 1, 2, 3, 5, and 7
Next Generation Science Standards Science and Engineering Practices 1, 2, 3, 6, and 8

Launch (Task presentation)

Introduce the lesson by asking students if they know an engineer (someone in their family or a friend). Show the three-minute YouTube video called *What Is an Engineer?* (which shows students from the University of Michigan defining engineering). Ask students to list characteristics of engineers, including types of engineers (mechanical, civil, etc.).

Tell students they are going to pretend that they are engineering students from the University of Michigan and their professor came up with a building competition. They will have 10 minutes to build the highest structure possible using only 15 pipe cleaners. Discuss the fact that engineers often have very limited materials (or funds) to make their designs, and even the plan itself is unknown at the start. With an engineering partner, they will make their structure. The teacher will measure the height of each structure from the table to the top of the structure. The pair with the highest structure is the winner of the competition.

Exploration (Anticipated student thinking—include class structure [in small groups, with partners, individually] and potential correct and incorrect strategies or solutions)

Student solution 1: Some students will begin immediately at random with no design anticipated beforehand.

Student solution 2: Some students will attempt to create a plan before building. They might discuss the base of the design, its shape, its size.

Student solution 3: Some students might use a circle for the base and create a cylinder-type shape for the base figure.

Student solution 4: Some students might use a triangular base or a square base.

Student solution 5: Some students might conclude that a circle is the strongest base, with a triangle second.

Four minutes into construction: Stop the students and tell them that sometimes engineers have funding cut mid-project or encounter other constraints that they did not anticipate. So, to simulate such a problem, they must now work with one arm behind their back.

Anticipation: Students will have to work more closely with one another, talking more and giving precise instructions.

Seven minutes into construction: Stop the students again and tell them that engineers sometimes work in global markets and must communicate with people who speak different languages. They must now work in silence without verbally communicating with their partner.

Anticipation: Students will have to use nonverbal cues to express their instructions. Having made a clear plan before construction will become an obvious support in this part of the lesson.

FIGURE 4.11

Partial Lesson Image for a Science Lesson on Skin Color (Part 1)

Science, Technology, Engineering, or Mathematics Goal(s):

The science goal for this lesson is to elicit students' informal knowledge about the causes of skin cancer and to begin to investigate, using mathematical data, the relationship between outside factors (e.g., UV light) and skin color.

Rationale: The rationale for this lesson is to set the context and hook the students with regard to the investigation into skin color, heredity, and the effect that the sun has on pigmentation.

State Standard(s): HS-LS2-2: Use mathematical representations to support and revise explanations based on evidence about factors affecting biodiversity and populations in ecosystems of different scales.

Cycle 1

Launch (Task presentation)

Relate a personal story about a family member or friend who was diagnosed with skin cancer. Ask students if they have known someone that has been treated for skin cancer. Talk through the scenarios the students share—for example, how the person knew or suspected he or she had a problem. Lead into the story below. Have students read silently and then do a "character-scene-plot" analysis to determine the meaning of the story.

"Stop it!" cried Tatiana.

Her dad, Dr. Disotell, was inspecting her skin very carefully.

"Look," he said, sounding serious. "Today a woman walked into my clinic for her annual physical exam. Everything about her seemed fine. She leads a balanced lifestyle, eats well, and exercises regularly. She's healthy! But as she was about to leave, I noticed a mole on her arm. It had many of the warning signs of skin cancer. So I removed the mole. This woman now has to wait for the lab results to see if it was cancerous. If it is, maybe we caught it early enough to treat it, but maybe not. Either way, her life is changed. I just want to make sure you don't have any suspicious moles, OK?"

Tatiana relented and allowed her dad to examine her skin. She asked, "Do only white people get skin cancer?"

"No, people of all skin tones can get skin cancer, but it does occur more frequently in Caucasians."

When the class has finished analyzing the story, have students get into groups of three, and ask them to discuss the following questions:

1. What are the causes of skin cancer?
2. At what age does skin cancer *typically* occur?
3. Do you think Caucasians are more at risk of skin cancer than other populations? Why?

Ask small groups to share their answers with the whole class. Record their answers.

- Possible responses to question 1: The sun, UV rays, not using sunscreen, heredity, tanning beds, chemicals (e.g., Agent Orange), foods, deodorant, mosquito spray
- Possible responses to question 2: 40s, old age (more sun exposure, time), any age, 60s
- Possible responses to question 3: They have lighter pigment, they like to tan (exposure), skin burns easier, DNA, melanin

In the launch of the first lesson, students hear a brief story about skin color and the relationship between melanin and UV light from the sun. If it arises in the unit launch, capitalizing on students' prior knowledge that sunshine plays a role in skin color is paramount. Students are then prompted to look at a map of the world to determine if the amount of exposure to UV light might affect skin color. Before reading our lesson image for the launch of the first task (Figure 4.12), hypothesize the solutions that students might give for the relationship between UV light exposure and latitude.

To anticipate students' solutions and explanations in this lesson image, we examined the graph as a small learning community of teachers and conjectured together about the diversity of arguments our students might make.

FIGURE 4.12

Partial Lesson Image for a Science Lesson on Skin Color (Part 2)

Launch (Task presentation)

Using the literacy strategy of your choice, have students read the first section of "Part II: Skin Pigmentation and UV Light," titled *Humans Were Initially Lightly Pigmented*.

Discuss the reading as a class. Answer any questions students have. Ask one or two students to restate the main ideas, and clarify any misunderstandings.

Have students read the second section, *Melanin: Natural Sunscreen*. Ask: ***Is there a relationship between the amount of UV light exposure and skin color?*** Introduce the image of the Global UV Index Forecast, and have students examine it in small groups.

Exploration (Anticipated student thinking—include class structure [in small groups, with partners, individually] and potential correct and incorrect strategies or solutions)

The closer you are to the equator, the more UV rays there are.

The equator is closest to the sun, so it gets more UV rays.

There is more sun near the equator.

Most sun is at 0 degrees latitude, with the least sun/UV rays at 90°.

Antarctica, northern Europe, northern Russia, and Canada have the least amount of exposure to sun, and so the lightest people. The darkest skin will be in central Africa, central South America, islands in Caribbean, and islands in the Pacific.

Lesson adaption suggested by Joanna Schimizzi.

Before Reading Chapter 5 . . .

In the next chapter, we examine what to do with the anticipations you've brainstormed—how to plan for whole-class discussions that are driven by students' thinking and capitalize on their work during explore time. Consider these questions before moving on:

• What kinds of questions do you ask during whole-class discussion?

• How do you determine what questions you will ask students during a whole-class discussion?

• How do you decide whom to call on during the discussion?

5

Imaging Mathematically Powerful Whole-Class Discussions

The lesson image process is like a rehearsal before the performance. It can help a teacher determine what to expect from student responses and how to create a whole-class discussion centered on students' mathematical ideas.

—Ashley Dickey, middle school mathematics teacher, Florida

In conversations with STEM teachers, the question most frequently asked of us is, "How do you know what questions to ask in a whole-class discussion?" We acknowledge that this segment of lesson implementation can be the most difficult activity of one's inquiry practice. However, the four important activities outlined in the previous chapters will make imaging your whole-class discussions much more powerful:

1. Unpack the STEM learning goals that are targeted for the lesson.
2. Choose appropriate tasks that allow for exploration of the goals.
3. Launch the task to engage students in the constraints and possibilities involved in problem solving.
4. Anticipate how students might solve the problem, both correctly and incorrectly.

Note that we did not say the process will be *easier*—but it will be more productive for you and your students.

This chapter reflects on these four imaging activities and illustrates how they can be used to envision a well-organized discussion, with students' solutions and explanations as the driving force.

Important Elements in Imaging a Whole-Class Discussion

One of the final components in lesson imaging is imagining the flow of the whole-class discussion so that the mathematical ideas identified at the beginning of the imaging process can be addressed. This process entails an intermingling of all the imaging from the previous chapters:

- What mathematical ideas are the activities designed to elicit?
- If students solve the problem in the way we anticipated, which solution strategies should we select, and in what order should they be presented, so that the mathematical ideas emerge?
- Which student representations would be most helpful for advancing the discussion, and how should we symbolize their thinking to aid in the discussion?
- What questions should we ask, and when should we interject them?

The answers to these questions will be different for every lesson and will also depend on prior work in the lesson imaging template.

Let us think through these questions with an example. Recall the problem involving ratios that was introduced in Chapter 2, where students had to determine whether there were enough food bars to feed a certain number of aliens. Students solve three problems as part of the first inquiry cycle. A 6th grade mathematics team has partially completed the lesson image template for the next lesson in the unit (Figure 5.1).

Since the goal of the lesson is to capitalize on students' symbolizing in order to introduce a ratio table, the teachers analyze the anticipated strategies with an eye to solutions that would lead toward this goal:

- While Solutions C and D are correct, they "hide" the linked composites (i.e., that there are three aliens fed by each food bar) and are too abstract for students who are just beginning to create these links.

FIGURE 5.1

Lesson Image for Cycle 2 of a Unit on Ratios

Science, Technology, Engineering, or Mathematics Goal(s): The idea of this lesson is to encourage students to link two composites together and to begin to organize these links when there are large quantities involved.

Rationale: Students need to find a way to organize the links as they increase in size. A ratio table should be introduced from students' work on this page.

State Standard(s): CCSS.Math.Content.6.RP.A.3: Use ratio and rate reasoning to solve real-world and mathematical problems, e.g., by reasoning about tables of equivalent ratios, tape diagrams, double number line diagrams, or equations.

Cycle 2

Launch (Task presentation)

[Note: 1 food bar feeds 3 aliens.]

1. Will 12 food bars be enough to feed 36 aliens? Explain.
2. Will 24 food bars be enough to feed 72 aliens? Explain.
3. Will 6 food bars be enough to feed 18 aliens? Explain.
4. Will 8 food bars be enough to feed 20 aliens? Explain.
5. How many food bars are needed to feed 39 aliens? Explain.

To launch this task, ask students what the picture above means. Tell them that their goal for the next 5–10 minutes is to determine the answers to the five questions—but most importantly, they should put some type of writing or drawing on their paper to show others in class how they found their answers. Use the Think-Pair-Share strategy by having students work independently for about three minutes, then let them know that they should work with their partner as soon as they are ready.

Exploration (Anticipated student thinking—include class structure [in small groups, with partners, individually] and potential correct and incorrect strategies or solutions)

Question 1 is our headliner. We expect the following solution strategies from students:

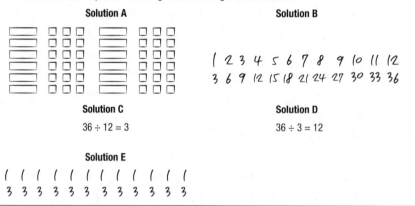

Whole-Class Discussion (Include tools, symbolizing, technologies, and questions you might pose)

• Solution A is the most concrete representation of the linked composites (one food bar for every three aliens) and might be a good one to begin with.

• Solutions B and E are both numerical representations of the linked composites, with Solution E showing the explicit link each time (1 to 3, written 12 times) and Solution B showing the accumulation of food bars and aliens (1 to 3, 2 to 6, etc.). The teachers decide that both solutions are productive and helpful because they are more efficient strategies than drawing 36 pictures. Solution B is particularly helpful for moving the mathematical agenda forward because it can lead to the introduction of a ratio table.

• The teachers decide that while Solutions C and D are both efficient and sophisticated, they will save those solutions for another discussion, after the ratio table has been introduced.

The teachers' final anticipated order of the solution strategies, then, is to begin with A, the picture, then follow up with E, and end with B.

When the teachers carry out the lesson plan in the classroom, they will monitor students' explorations and record on their tablets which students create the strategies they anticipated. It is then up to each teacher to decide which students who used Solutions A, B, and E will present their solutions to the class. With the decision made about which solutions will be presented and in which order, the teachers then begin to image how the whole-class discussion will flow.

It is helpful to have all three solution strategies on the board at one time for comparing and contrasting. Hence, about six minutes into exploration time, the teacher will have collected data from the students to determine who will present which strategy; student presenters are then invited to inscribe their solution on the board while the rest of the class finishes the problems. Having students do this toward the end of exploration saves time *and* is a smart classroom management strategy. If the teacher waited until everyone was finished with all the problems and *then* asked students to write their strategies on the board, the rest of the class would have to sit and wait, with no mathematical activity to work on, which can lead to discipline issues.

Instead, when the teacher calls the class to order to discuss their solutions, the strategies are already on the board, and no additional time is wasted.

It is also important to have the students explain their strategy to the class. The teacher will pause after Solution A is presented and ask the class if there are any questions for the student presenter. This allows students the opportunity to ask clarifying questions. After the explanations for Solutions E and B are presented, the teacher will do the same thing. The teacher will then ask students to take two minutes to decide with their partner which of the three strategies is the most efficient. The teachers predict that students will think that Solution B is the most efficient, with E second and A third.

Why do the teachers ask students to determine which solution is most efficient? Recall that a goal of this lesson is to introduce a ratio table based on student reasoning. Students probably would not organize their work in a ratio table naturally, unless they had studied tables in a previous grade. However, Solution B is a way that students would naturally organize the quantities involved in this problem, so the teachers imagine capitalizing on this student-generated strategy to introduce the ratio table—a mathematical representation that will help students organize and solve future problems more efficiently. The teachers envision saying something like, "So, most of you decided that Solution B was the quickest way to solve the problem. I want to show you how mathematicians might write it. They would put a horizontal line to separate the food bars and aliens and even label them to remember what the numbers represent. Then they would make a table by putting vertical lines to separate the numbers each time they increase" (Figure 5.2).

FIGURE 5.2

Teachers' Imaged Representation for Helping Students Structure Ratios in a Table

Food bars	1	2	3	4	5	6	7	8	9	10	11	12
Aliens	3	6	9	12	15	18	21	24	27	30	33	36

The teachers end this second cycle of inquiry by deciding to pose a new task with a different food bar-to-alien ratio and asking students to draw either a picture or a ratio table to show their reasoning. They will collect students' independently constructed responses and use them to determine what sense each student made from the discussion.

Know Your Lesson Goal

A common misunderstanding about inquiry teaching is that "anything goes"—whole-class discussions are open-ended, and anyone can say anything. In fact, quite the opposite should be true. The main goal in lesson imaging the whole-class discussion is to ensure that the contributions made by students and the teacher move the lesson forward in a crisp, mathematically sound fashion. This requires precise imaging and control during the planning and implementation of the lessons. As depicted in the example above, once the teachers have anticipated the variety of solutions the students might create, they must then decide which ones to capitalize on in class.

Additionally, not all solution methods need to be shared during the whole-class discussion. Remember that the teachers in the ratio example above decided to postpone sharing of the more abstract division strategies because the methods did not fit their agenda for that day. Since the goal was to use students' strategies to introduce a ratio table, they decided that the division methods did not provide natural opportunities for the ratio table to emerge.

Do Not "Fix" Mistakes During Explore Time

There are times when wrong answers arise quite naturally during the problem-solving process. Take, for example, a ratio task posed several days after a shortened ratio table had been introduced (Figure 5.3).

The teacher could logically anticipate that some students will say the answer is 9, especially if this task is early in the unit. Students might look at the 1 and the 4 and reason that since 3 teachers were added, they have to add 3 toddlers to the original 6, which gives them 9. This error is especially common when students have been taught in previous grades or units that *whatever you do to one side, you do to the other*. This type of reasoning is called *additive*, and it's very common as students first begin to reason about ratios and proportions.

FIGURE 5.3

A Task Using a Ratio Table

At Tiny Tots Day Care, the ratio of teachers to toddlers must always be 1 to 6 in the classroom. How many toddlers can be in the room if there are 4 teachers?

Teachers	1	4
Toddlers	6	?

Wrong answers can move your mathematical agenda forward. Rather than "fix" each student's incorrect strategy during explore time, it is important to share it with the whole class so that all students are confronted with the method, can analyze it for correctness (or not), and can then create a meaningful reason that it is mathematically incorrect.

When we have had this discussion in our classes, students have evoked linked composites as a rationale for why the answer "9" is incorrect. Students argue that if each teacher has 6 toddlers, 2 will have 12, 3 will have 18, and 4 will have 24. If there were only 9 toddlers, you wouldn't need that many teachers. They often draw a picture to prove why the link is broken (Figure 5.4).

Such a conversation is important to bring back the notion of a multiplicative link between teachers and toddlers, which might be challenged in this new context.

The key point here is that mistakes are important opportunities for learning—not just for the person who gets the incorrect answer, but also for other students in class who might need to revisit the mathematical idea.

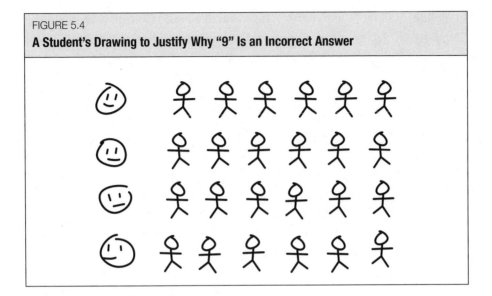

FIGURE 5.4

A Student's Drawing to Justify Why "9" Is an Incorrect Answer

Start with Different Answers on the Board

A technique that is often used to motivate genuine student engagement is to begin the whole-class discussion with two or three different student-derived answers on the board. The question to the students can then be, "Which solution or solutions are correct?" This accomplishes two things:

• It allows for the possibility that more than one answer is correct—and for some mathematical activities, that is true. Many students have been led to believe that there is only one right answer for all mathematics problems. For example, we have used problems such as the one in Figure 5.5 in 1st and 2nd grade to encourage students' arithmetical reasoning and understanding that some problems have multiple correct solutions.

• If the correct answer is unknown, students are more invested in seeing if their own answer is correct. Students will work diligently to determine a strong explanation that proves their answer is correct, or they will discover a mistake in their reasoning. This self-evaluation and critique of other students' reasoning is invaluable for learning mathematics and also constitutes one of the Standards for Mathematical Practice in the Common Core State Standards for Mathematics (National Governors Association Center for Best Practices & Council of Chief State School Officers, 2010).

Not every mathematics discussion allows for opening with a debate, but when it can, the arguments can be very strong and the students are motivated to explore important mathematics.

FIGURE 5.5

An Activity That Has More Than One Correct Answer

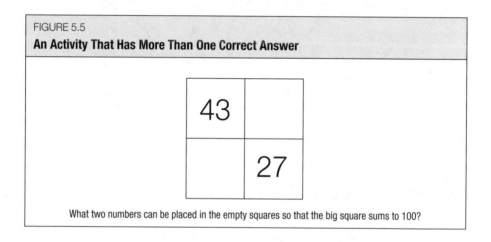

What two numbers can be placed in the empty squares so that the big square sums to 100?

Do Not Do the Comparing and Contrasting for Your Students

Once students' solutions are on the board and they have each explained their reasoning, the most important part of the lesson begins. Teachers generally know that the students must "see" the differences and evaluate the solutions. However, the teacher is usually the person who reveals those differences or similarities. The teacher might say, "I want you to examine these three solutions and see what is the same and what is different among the three." Then, after about 30 seconds, the teacher begins to name the similarities and differences. The teacher is the one who validates the most efficient or sophisticated solution.

Instead, we stress that the teacher should ask the question and give students a few minutes with their partners to make those comparisons or determinations for themselves. Pairs should then present the similarities and differences they found to their classmates.

Don't Just Ask "Why?" or "How Did You Do That?"

While these are two very important questions, they can also be very vague for students. Let's return to the problem in Figure 5.3, in which a student incorrectly added three toddlers to the six. The teacher might follow up by asking, "How did you get nine as your answer?" The student's reply is simple: she added three more toddlers. The teacher's next question might be, "Why did you do that?" to which the student says, "Because I added three more teachers." A typical follow-up at this point is to ask more questions that actually *tell a student how to think*, rather than allow the student to understand why his or her thinking is invalid.

Consider the follow-up questioning below:

Teacher: OK, so you added three more toddlers because you added three more teachers. But how many teachers do there need to be with six toddlers?
Student: One.
The teacher draws a picture of one teacher and six toddlers.
Teacher: So, how many toddlers would there be for two teachers?
The teacher draws one more teacher and six more toddlers.
Student: Twelve.
Teacher: Yes! Do you see now how it is 24 toddlers for 4 teachers?
Student: Yes.

In this exchange, the teacher genuinely tries to help the student, yet the teacher is the person doing the problem solving—attempting to "fix" the student's thinking through the questions being posed. Questioning techniques that begin open and tend to become more closed, with the answer to the final question basically answered by the teachers themselves, resemble a funnel pattern (Herbel-Eisenmann & Breyfogle, 2005; Voigt, 1995) (Figure 5.6).

Instead, the teacher should choose that student to present her strategy in whole-class discussion and ask her to prove it. Other students will disagree and will provide pictures to contradict her strategy. This allows the student to find a more visual way to defend her thinking and also allows her to analyze and compare her rationale with her peers' strategies. It ultimately gives the student and her peers authority over their own problem solving and validation of their methods, rather than depending solely on the teacher.

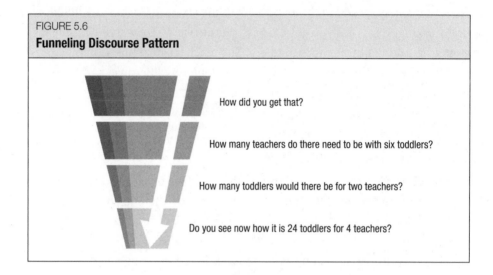

FIGURE 5.6
Funneling Discourse Pattern

How did you get that?

How many teachers do there need to be with six toddlers?

How many toddlers would there be for two teachers?

Do you see now how it is 24 toddlers for 4 teachers?

Use Social Norms to Guide Your Selection of Students to Present

During the lesson imaging, teachers decided which strategies should be presented and in what order. The next step is to decide which students will speak. This can be determined by thinking about *social norms*, the expectations that teachers and students have for one another when participating in public discussions (Stephan & Whitenack, 2003; Yackel & Cobb, 1996). For example,

- Students are expected to explain their reasoning.
- Students are expected to understand one another's strategies.
- Students are expected to ask clarifying questions when they do not understand.
- Students are expected to indicate agreement or disagreement.

Say, for instance, that Kimberly has not been sharing her thinking lately, but she has drawn a picture of 8 food bars and 24 aliens. David, who talks a lot in class, has the same solution. The teacher may decide to have Kimberly present rather than David, so that Kimberly both understands and meets the expectation that she is supposed to share her thinking in class.

Perhaps another student shares a solution that is very sophisticated, and the teacher sees a confused look on Deondre's face, but he does not have his hand raised. This is a good opportunity to call on Deondre and give him an opportunity to ask his question, reminding him of his responsibility to do so.

The portion of class where the teacher asks students to work with their partners to determine the similarities and differences among the strategies illustrates the social norm that the teacher expects students to create viable arguments, in collaboration with peers. From a classroom management perspective, it also avoids that quiet time in class where no one wants to answer the question, possibly because the students haven't had a chance to process their thinking yet.

Ask Higher-Order Thinking Questions

Another type of question that can be asked to further a whole-class discussion is one that involves *sociomathematical norms*—the criteria for what counts as mathematical participation in class. These norms are negotiated among teachers and students (Stephan & Whitenack, 2003; Yackel & Cobb, 1996). For example,

- What counts as an efficient mathematical solution?
- What counts as a different mathematical solution?
- What counts as a sophisticated mathematical solution?
- What counts as an acceptable mathematical explanation?

Recall the ratio discussion above, when the teacher asked which of the three solution methods was the most efficient. This question was imperative here, because comparing the strategies for efficiency required students to analyze them on a higher level (Bloom, 1972). The question also provided students with the opportunity to make contributions that furthered the teacher's mathematical agenda. Judging the efficiency and sophistication of different methods and the differences among methods is an important mathematical practice because it leads to more advanced mathematical thinking.

What counts as an acceptable mathematical explanation? can lead to another important set of questions from the teacher. Thompson, Philipp, Thompson, and Boyd (1994) suggest that there are two types of orientations that students and teachers can have toward mathematics:

• Those who think that mathematics is about getting answers and calculating by following a prescribed set of steps have a *calculational orientation* to mathematics. A teacher who accepts a procedure by itself as an adequate explanation would be seen as having a calculational orientation.

• Those who view mathematics as both procedures and understanding the "why" behind those procedures have a *conceptual orientation* to mathematics. A teacher with this orientation might follow up a student's procedural explanation with questions such as, "Why did you add there?" or "When you divided those two numbers and got 35, what does that 35 stand for in the story?" These types of questions go beyond just stating the calculations that were used to solve the problem by asking students to provide their rationale for the procedure and what the numbers mean in terms of the original problem.

According to Thompson and colleagues (1994), furnishing the reasons behind the calculations is critical for those students who did not use the same calculations or who had a difficult time solving the problem. Simply giving students the steps does not provide mathematical meaning and does not help struggling students interpret future problems. With conceptual support, struggling students have a better chance at applying the strategies to new situations.

Call on the Most Visual Response, but Not Because It's the Most Concrete

When deciding in what order to have students present their methods, it is often tempting to begin all-class discussions with the most concrete or visual strategy. While we acknowledge that most times this makes sense, it is probably not for the reason you might presume. It is true that visual methods are easier for other students to "see" or interpret; however, we propose that the power of visual aids during whole-class discussion lies in supporting students' ability to understand the meaning behind others' methods.

For example, when solving the problem of how many food bars feed 30 aliens, given a 2:3 food bar-to-alien ratio, we have seen students use an abbreviated ratio table (Figure 5.7).

FIGURE 5.7
Abbreviated Ratio Table

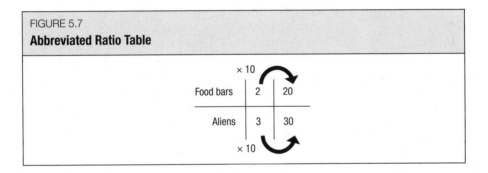

When we ask why they multiplied by 10, the most common response we get is, "It is what gets you to 30, so you should multiply 2 by 10 also." This response focuses solely on the procedure; it does not address why multiplication is involved at all.

But consider the difference if one of the many solutions on the board resembled Figure 5.8. Students can now more easily determine the meaning of "× 10," explaining that the top "× 10" in the ratio table signifies 10 groups of 2 food bars, and the bottom "× 10" stands for the 10 groups of 3 aliens. Without this firm foundation of meaning, the ratio table can simply be a calculational tool with little meaning. The concrete, visual solution strategy provides the class with an opportunity to make conceptual sense of the calculations in the ratio table.

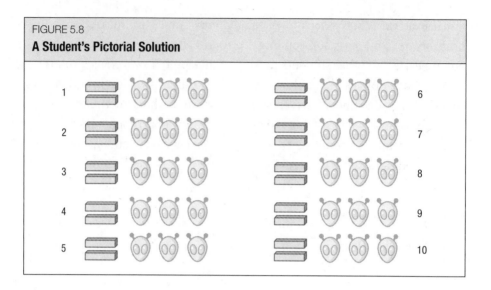

FIGURE 5.8

A Student's Pictorial Solution

An Engineering-Mathematics Example

Let's explore one more example of lesson imaging a whole-class discussion. After an engineering activity on skyscraper construction,[1] the teacher asks 9th graders to pretend they are the supervisors of a construction company that has been hired to take over the construction of the world's newest skyscraper (Underwood, 2002). The launch takes about 20 minutes. The students are then given three problems to solve (Figure 5.9).

For problem 1, the teachers anticipated that some students would need to draw every building for each week until they reached the 11th week, keeping track of the height of the building each time, as shown in Solution A (Figure 5.10).

Another way students might solve this problem is to simply draw the tops of the buildings each time, keeping track of the height and week

---

[1]The text of this lesson, Skyscraper Basics, can be found on WGBH's Building Big website (www.pbs.org/wgbh/buildingbig/skyscraper/basics.html).

FIGURE 5.9

First Activity Page of Engineering-Mathematics Unit on Slope

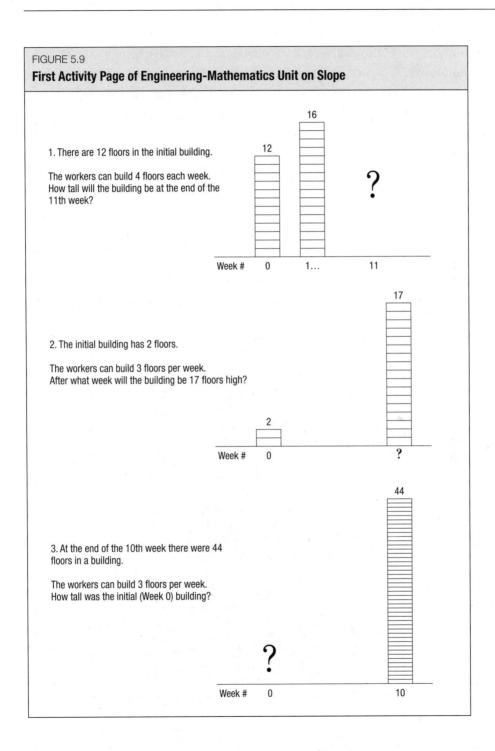

1. There are 12 floors in the initial building.

The workers can build 4 floors each week. How tall will the building be at the end of the 11th week?

2. The initial building has 2 floors.

The workers can build 3 floors per week. After what week will the building be 17 floors high?

3. At the end of the 10th week there were 44 floors in a building.

The workers can build 3 floors per week. How tall was the initial (Week 0) building?

number simultaneously, as shown in Solution B. Finally, some students may find keeping track of each building cumbersome and simply notice that there are 10 weeks between Weeks 1 and 11; that, they conclude, must equal a total of 40 new floors (10 weeks at 4 floors per week). Students who use this strategy may remember to add 40 to 16, or they may add 40 to 12, or they may forget to add 40 to anything, for the answers of 56, 52, and 40, respectively.

Before reading further, try to decide which of the three strategies and answers you would make sure were presented and in which order.

FIGURE 5.10

Anticipated Solutions for Problem 1

Exploration (Anticipated student thinking—include class structure [in small groups, with partners, individually] and potential correct and incorrect strategies or solutions)

All three questions are important, so ask students to solve the first two first.

We expect the following solution strategies from students for problem 1:

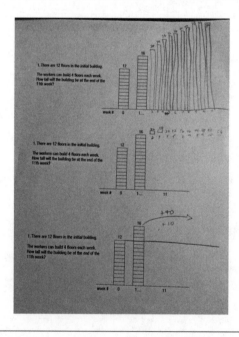

Our vision is shown in Figure 5.11.

Notice that the teachers decided to begin by listing all possible answers from students as a way to motivate students to get involved in the discussion. The teachers would then encourage students to commit to a solution and be ready to defend it.

FIGURE 5.11
Imaged Order of Presented Strategies

Whole-Class Discussion (Include tools, symbolizing, technologies, and questions you might pose)

Assume that all three strategies (and three different answers) are present. Ask students for their answers, and put 56, 52, and 40 on the board as three possibilities. Ask students which one they think is correct. There should be various opinions, but 56 will be the most popular. Ask a student with Solution A and one with Solution B to put their pictures on the board and explain them. Then ask students with pictures similar to Solution C but different answers to explain their solutions.

While students are explaining Solution C, encourage the class to ask questions about how the presenters created their solution. If the students do not ask, make sure to point out that all those who used Solution C got 40 as part of their answer, but what does the 40 stand for in the problem? (Students should point out that those are the extra floors.) The teacher can symbolize the "extra floors" by drawing a line across the top of the original building and putting "+40" if the presenters did not do so themselves—or the teacher can ask a presenter to show where the 40 extra floors are. Make sure that there is closure from the students regarding which number to add to 40 and why. Students and the teacher can refer to Solutions A and B to help visualize the 40 floors.

Next, the teachers decide to start with the most visual and least sophisticated strategy as a way to form a strong foundation for the remaining contributions. Solution B is simply a curtailment of Solution A, but it might lead students to a table-type structure in the future. If a student creates a table, the teacher would have that student present next, to match the picture in Solution B.

Finally, the teachers brainstorm some important questions and symbolizing techniques for Solution C. After students explain their three different answers, the teacher can pose the question, "They all found +40! What does 40 stand for?" This type of question acknowledges that 40 is an important quantity for this problem and can even be found in the drawings, if necessary.

In terms of the mathematical agenda for the day, the teachers know that they want to help students structure linear growth patterns as an initial amount (y-intercept) plus a rate times amount of time (x). Structuring the empty space between the initial building's height and its height at Week 11 as 10 rates of 4 per week will be critical for future problems. The line drawn by either the students or the teachers can be an important representational support to visually structure the empty space between the top of the initial building and the 11th week. What lies under the line is the starting amount, which will later be named y-intercept when the students start placing building heights on a Cartesian plane. Focusing the discourse on the meaning of the numbers in the problem and symbolizing with a line are two important tools for the teacher to promote the meaningful flow of the discussion.

After concluding the discussion of the first problem, the teacher can guide the discourse for problem 2 in a very similar way (Figure 5.12). Notice the thin lines either drawn or introduced by the teacher in Solution A^2 and the line drawn in Solution B^2. While the teachers agree that Solution A^2 will come first, they cannot decide whether to go with Solution B^2 next and then C^2, or C^2 and then B^2. If C^2 (subtracting three at a time from the top of the last building until you reach the initial building of two floors) is explained first, then the student who explains Solution B^2 can use Solution C^2 to explain what the "15" in the gap stands for and why the student divided 15 by 3. On the other hand, if the class starts with B^2, students who have difficulty understanding why the division step was conducted can then consider Solution C^2, which provides visual support for understanding Solution B^2.

The teachers conclude that it doesn't matter which comes after A^2 as long as Solutions B^2 and C^2 are used to support a conceptual discussion of linear growth. At the conclusion of the discussion of problem 2, the teacher will ask students to use what they have learned to complete problem 3. In Figure 5.13, we present the anticipated student reasoning and potential whole-class discussion that may arise from problem 3.

Again, the teachers decide to begin the discussion by getting all answers on the board to motivate students to analyze the reasoning of others. They decide to begin with the least sophisticated first to establish a baseline answer as correct until someone can prove it wrong. The question "Who has an easier way?" is motivated by the teacher's ability to insert higher-order questions (the sociomathematical norm of "what counts as an efficient solution") at the right time to prompt more sophisticated strategies and to provide support for those students who only have "guess and check" at their disposal. The strategies the teachers envision move from pictorial support to more abstract solutions ($44 - 30 = 14$). The representational support is presented by students first, which helps the class make sense of the more abstract calculations (i.e., they can have a discussion that is conceptual and not just procedural).

FIGURE 5.12
Anticipated Solutions for Problem 2

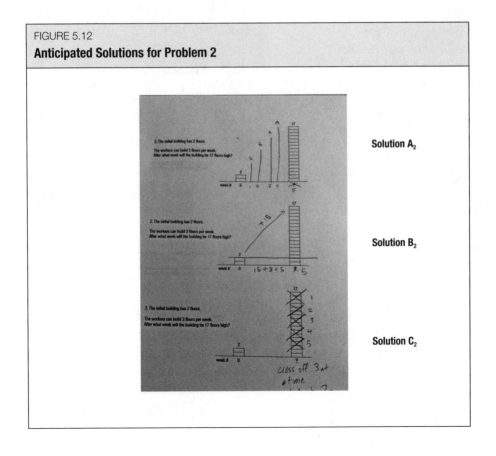

FIGURE 5.13

Visual Support Created for Problem 3

Exploration (Anticipated student thinking—include class structure [in small groups, with partners, individually] and potential correct and incorrect strategies or solutions)

A. Most solutions will involve pictures of the type here. Some students might use circles instead of lines to indicate taking off 3 floors at a time and stopping when they reach 0 weeks (10 altogether). They will then count the remaining floors to get 14 floors.

B. Some students may stop when they reach 1 to get 17 floors.

C. Some students will guess an initial height at Week 0 and then check to see if they get to 44 when adding 3 each time.

D. Some students will know that there is a total of 30 floors built over 10 weeks, so they subtract those 30 off the top of the 44 to get 14 for Week 0.

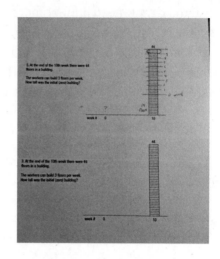

Whole-Class Discussion (Include tools, symbolizing, technologies, and questions you might pose)

Start with putting different answers on the board (possibly 14 and 17, maybe others). Guess-and-check person can begin the discussion, and the teacher can ask if anyone had an easier way. Teacher will call on students who get Solutions A and B (above).

Make sure that we highlight the numbers to the right-hand side of the tall tower. What do those numbers stand for? Make sure that students get the idea that the numbers keep track of the weeks of three floors. There should be 10 of those sets of 3 on top of the initial building. Teacher can draw the "initial" line again, starting at the top of the 14th floor on Week 10 and going to the beginning.

Teacher then asks Solution D author to explain. Have students relate the numbers in Solution D (i.e., 30, 10, and 14) to the picture for Solution B. Students should see that the 30 in Solution D is 30 total floors cut into 10 sets of 3 in the picture for Solution B.

Teacher should summarize by showing the following picture for the 10th week:

The gray lines indicate the top of the building at the end of that number of weeks.

10 weeks x 3 floors per week = 30 total floors.

The teachers decide to encourage symbolizing that breaks the last tower into chunks of three (the rate) to help students see that there will be w number of rates in the wth week sitting on top of an initial amount of floors. In that way, the teachers are supporting (and laying the groundwork for) a visually and quantitatively meaningful structuring of the formula $y = mx + b$. In fact, not long after this part of the unit, the teachers will ask students to write an equation for the building pattern, and students will quite naturally write formulas such as $T = 3w + 14$ for the example above.

Conclusion

In this chapter, we brought the most critical piece of lesson imaging to the forefront: the whole-class discussion. It is in this segment that mathematical connections, representations, and higher-order conclusions should emerge in conversation. As demonstrated in this chapter, these connections do not happen by magic—they do not come from students sharing in a show-and-tell fashion how they created their solutions. Instead, students' anticipated solution strategies should be used to envision the flow of a whole-class discussion that best supports students in the mathematical growth that was intended from the outset.

The teacher's role in the whole-class discussion is to ensure the following:

• The established social norms are reinforced: students are expected to share their reasoning, analyze the reasoning of others, ask clarifying questions, and indicate agreement or disagreement.

• Only solution strategies that contribute to the mathematical goals of the lesson and that emerge in discourse are shared.

• The order of the presentations from students best facilitates student analysis and sense-making of the mathematical ideas.

• Higher-order questions are asked that promote conceptual discourse and focus on more than the procedures that are used (e.g., "Why did you divide?" "What does your answer mean in the building pattern?").

• Comparing and contrasting of solution strategies is done by *students*, not the teacher.

• Supportive visual and procedural representations are used to provide students with the ability to structure and organize their thinking and to make sense of their own and others' strategies.

Before Reading Chapter 6...

Consider these questions before moving on to the next chapter:

• What questions do you still have about lesson imaging?
• Which parts of the process do you think will be hardest for you?
• How do you enact a well-designed lesson image?

6

Putting It All Together

Unpacking the learning goal, working out the tasks, and anticipating student responses occur somewhat simultaneously. We come to our planning meeting with a mathematical big idea and a task. We start with all of us working out the problem (keeping our students in mind) and sharing our answers and strategies, and an organic discussion ensues. This discussion often leads us to refine our goal, think of even more diverse student responses, design an effective launch, and imagine how the whole-class discussion would flow. This teacher discourse covered most of the lesson imaging tenets, not in a linear way, but in a more fluid and natural way.

—George McManus, middle school mathematics teacher, Florida

In the preceding chapters, we presented a number of strategies for lesson imaging that draw on students' diverse problem-solving abilities. We began by discussing the commonalities that cut across the STEM disciplines, focusing on modeling and the scientific method. We acknowledged that while lesson imaging can be used to plan lectures, it is most useful for instructional methods that use the *teaching for autonomy* approach, in which problems are posed and students are required to create personally meaningful strategies, rather than have a strategy modeled for them by the teacher. When the whole-class discussion is dependent on the students' problem-solving activity, the flow of the lesson is much less predictable than when lecturing. Thus, lesson imaging provides a structure for teachers to imagine how students might solve problems and how teachers might engineer the follow-up whole-class discussion to ensure that the STEM objective is met.

We then introduced our lesson imaging template, which is being used by teachers across the country to plan for more powerful classroom discussions. Subsequent chapters detailed each aspect of this template: unpacking the STEM objective, choosing worthwhile tasks, launching the task, anticipating student reasoning, and engineering the whole-class debrief session.

In this chapter, we explore in depth a lesson that was used in a 7th grade classroom. We present each component of the lesson image and include examples of how the classroom session actually played out.

Lesson Imaging, from Launch to Assessment

The lesson image in this chapter was prepared for a 7th grade inclusion classroom of 25 students, 8 of whom have special needs. One author of this book, Julie Cline, was the regular education teacher, who co-taught with special educator Erika Allred. The two had been co-teaching and lesson imaging together for three years. While each teacher had specific students on her roster, they both considered each other's students their own and attempted to teach in a genuine co-teaching manner (Dieker, 2001). They co-imaged together on an almost daily basis and co-taught every day, with each teacher leading a cycle of tasks. For example, during explore time, both teachers monitored students' reasoning, kept track of strategies on their tablets, and huddled together before the whole-class discussion in order to share what they found and plan for the order of presentations. Ms. Cline might then lead the whole-class discussion, with Ms. Allred interjecting questions or comments. During the second cycle of launch-explore-summarize, Ms. Allred would lead the whole-class discussion, with Ms. Cline serving as the secondary teacher.

It is important to point out that huddling to compare notes and then sharing responsibility for whole-class discussions makes this a true co-teaching approach, as opposed to the mathematics teacher taking the lead while the special educator merely serves as the disciplinarian.

The class had been working on state standard CCSS.Math. Content.7.EE.B.4.a:

Solve word problems leading to equations of the form px + q = r *and* p(x + q) = r, *where* p, q, *and* r *are specific rational numbers. Solve equations of these forms fluently. Compare an algebraic solution to an arithmetic solution, identifying the sequence of the operations used in each approach.* (National Governors Association Center for Best Practices & Council of Chief State School Officers, 2010)

The teachers were using a unit from *Mathematics in Context* (Romberg & de Lange, 1998) on writing equations from visual patterns; students were asked to write equations for the tile patterns of different floors (as shown in Figure 0.1 in the Introduction). After completing this investigation, the teachers wanted students to take what they had learned about recursive equations (also known as Next/Now equations) and about writing more direct formulas for linear and nonlinear patterns, and apply that knowledge to creating linear formulas. To do this, the teachers adapted another investigation from *Mathematics in Context*, one that prompts students to create formulas based on bird patterns (Figure 6.1).

Before we show you the lesson image template that Ms. Cline and Ms. Allred created for this lesson, we suggest that you fill out the template yourself. Our advice is to work through the three problems yourself, including drawing the three patterns, even if you do not think you need to. When you have finished working through the problems yourself, then go back and attempt to solve the same problems in multiple ways. Once you've done that, go to the top of the template and write what you think the goal of the lesson is. Then, think about the launch of the activity and how students might solve the problems. Finally, think about the whole-class discussion: which solutions would you have students present and in what order, and which problems would you start the discussion with? Although the lesson imaging template has an implied, linear order (launch to assessment), discussion need not be so linear. Even if you are working with a teacher's manual that lists the academic goals of the lesson, we have found it more beneficial for teachers to work through the problems in order to "unpack" what this goal really means, especially from the student perspective. Therefore, most teachers we have worked with begin the lesson imaging process by listing the mathematical goals and then working through the open-ended task to get a sense of the deep mathematics or science embedded in the goal. Once teachers have a strong sense of the mathematics or science of the lesson, they continue the imaging process in a more dynamic way, addressing each component as the need or opportunity arises.

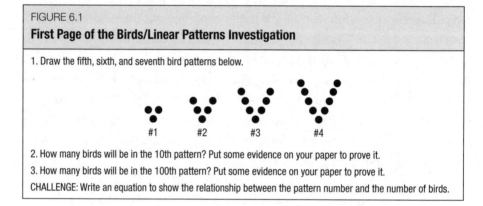

FIGURE 6.1

First Page of the Birds/Linear Patterns Investigation

1. Draw the fifth, sixth, and seventh bird patterns below.

#1   #2   #3   #4

2. How many birds will be in the 10th pattern? Put some evidence on your paper to prove it.

3. How many birds will be in the 100th pattern? Put some evidence on your paper to prove it.

CHALLENGE: Write an equation to show the relationship between the pattern number and the number of birds.

Anticipated Student Reasoning

To begin the meeting, Ms. Cline and Ms. Allred typically start by working through the problems (see Figure 6.1) from their own and their students' perspective. They conjectured a number of strategies during their co-planning session (Figure 6.2).

After working through the problem and anticipating a number of ways to solve it, Ms. Allred and Ms. Cline now more deeply understand the goal of the lesson and the subsequent problems in the investigation (Figure 6.3).

Ms. Cline and Ms. Allred also wanted to use the bird pattern investigation as an opportunity to reestablish the norm of cooperation in their class. With these goals in mind, the teachers revisited their anticipated student reasoning to determine which solutions to use in class and in what order.

FIGURE 6.2

Anticipated Student Reasoning for the First Page of the Birds Linear Investigation

Exploration (Anticipated student thinking—include class structure [in small groups, with partners, individually] and potential correct and incorrect strategies or solutions)

While students explore, we will walk around monitoring both their on-task behavior and the strategies or equations they create. We expect there to be, at a minimum, four different strategies:

3   3
Leader Bird

3
2
1
Wi-Fi

4   3
Plus One

- Solution 1 (Red): Identify one "leader bird" at the bottom and two "arms" of P dots
- Solution 2 (Green): Identify one "leader bird" at the bottom and P pairs of dots growing above
- Solution 3 (Blue): Identify $P + 1$ dots on the left arm and P dots on the right (no leader bird)
- Solution 4: Double the number of dots from the 5th pattern to get the 10th pattern
- Solution 5: Make a table of values from 1 to 10
- Solution 6: Use recursive reasoning: just keep adding twos until you get to the pattern you want
- Solution 7: Draw all the pictures to get the 10th pattern

We will not correct anyone's mistakes unless they miscount a dot.

100th pattern

Students who have Solutions 1–4 will be able to find the number of dots in the 100th pattern either by looking at their spatial structuring or by continuing the table:

Solution 1: $100 + 100 + 1$

Solution 2: $100 \times 2 + 1$

Solution 3: $101 + 100$

Solutions 4 and 5: Might continue the table or can erroneously take the 10th pattern (21) and multiply it by 10 to get 210

Solutions 6 and 7: Won't be able to get the answer

Formula

For those who can find a formula:

Solution 1: $2P + 1 = T$

Solution 2: $P \times 2 + 1 = T$

Solution 3: $(P + 1) + P = T$

Solution 6: $Next = Now + 2$

Before reading further, look back at the teachers' anticipated student reasoning (Figure 6.2) and determine (1) where you would start the whole-class discussion, (2) the order in which you would discuss the problems, and (3) the order of the presentations. Keep in mind the symbolizing you might use and the questions you might ask to ensure meeting the lesson objective.

FIGURE 6.3

Mathematical Goal of the Birds/Linear Patterns Investigation

Science, Technology, Engineering, or Mathematics Goal(s):

The objective of this lesson is for students to begin to understand that a linear equation is structured by a fixed amount and a constant rate. Students will be able to identify the constant rate of change from a spatial pattern and to write an equation that models that pattern.

In kid-speak: Students will learn that there is a leader bird (or a set of leader birds) that stays the same throughout the pattern and that the rest of the pattern can be structured into sets of the pattern number. This will eventually translate to the constant and the rate, especially if students relate their work to the Next/Now equations.

Revisit the class goal of helping one another grow in mathematical understanding, not competing with and trying to outdo one another.

NOTE: **Prior Knowledge:** Students have already studied variables as letters that represent unknown quantities with *Mathematics in Context* tiling problems.

The teachers' lesson image of the whole-class discussion is shown in Figure 6.4.

Now that the teachers have new ideas about the goal of the lesson and the types of reasoning that might emerge from students, they then turn to the task of imaging the launch. Working through the problem first helped them know that they should not reveal the plus-two pattern in the launch, as that is crucial to students developing a solution. Are there other insights you developed from anticipating students' thinking that you believe are important for structuring the launch? Before reading further, how would you launch this activity in the classroom? What contextual features are important to discuss? How can you launch this investigation in a way that engages students but does not steal their autonomy? Prepare your own launch, and compare it to the teachers' (Figure 6.5).

The teachers' final decision is how to assess students' learning. They decide to use an activity sheet that will reveal students' understanding of the mathematical objective for the day, given either as an exit slip or a home-work assignment (Figure 6.6). The teachers can then determine how many students are able to write formulas for spatial patterns, what the most pop-ular ways of reasoning are, if students are still stuck in the Next/Now equa-tions, and which students are still unable to write a formula.

FIGURE 6.4

Teachers' Image of the Whole-Class Discussion

Whole-Class Discussion (Include tools, symbolizing, technologies, and questions you might pose)

Toward the end of explore time, we will select certain students who have the seven different strategies to write their reasoning on the board. We will name each strategy.

Start the discussion by asking each student to tell the class how many birds are in the 10th pattern (e.g., 21 and 22). Students explain their method in the following order:

1. Solution 7 (drawing)
We will then ask if anyone found a more efficient way.
2. Solution 6 (recursive)
3. Solution 1 (two arms, leader bird)
4. Solution 2 (groups of two, leader bird)
5. Solution 3 (no leader bird, two arms)
6. Solutions 4 and 5

Small-Group Discussion (approximately 3 minutes)

Ask students to look at all the solutions on the board to determine their accuracy. They should copy the name of the person who authored each strategy and place one of three faces next to the name: smiley face indicates that you agree with the answer and understand the strategy, frowning face means that you disagree with the answer and strategy, and confused face means that you are not sure or do not understand the solution strategy.

Whole-Group Discussion (approximately 15–20 minutes)

After the small-group discussion, some students might want to take their solutions off the board because they no longer agree (such as finding the fifth pattern and then doubling it). Do not take it off, but make a note on the board that we will explore why they now think it doesn't work.

Ask the class how many students agree with Solutions 6 and 7.

Ask: "Which drawings on the board make it easier to see the answer?" Students will probably say Solutions 1–3. Have students summarize how the authors structured the patterns for each solution (1–3). Either the term "leader bird" will either come from students, or we can name it. Reiterate that Solution 1 is like two arms on a V with a leader bird (name it the arm method), Solution 2 is like a tornado (some students might name it the Wi-Fi method), and Solution 3 is the arm method with no leader bird.

Ask students to look at the table solutions, if those come up. Ask: "Why doesn't 'double the fifth pattern' work?" If necessary, support students' discussion by drawing a five-bird pattern on the board and then another five-bird pattern next to it as a "double." Students will notice that the leader bird gets copied twice, so they just need to subtract 1 from the 22.

Ask students to draw a quick sketch of what the 1,000th bird pattern would look like, and have someone draw their sketch on the board. Ask: "What would the Pth pattern look like?" (P's where the 100 would be, or $P + 1$ where the 101 is).

100th Pattern Discussion

Ask students to use at least two of the solution strategies on the board to find the number of birds in the 100th pattern. Have students present a couple, getting some variety on the board. Ask: "Why didn't anyone choose Solutions 5–7?" Discuss efficiency—some strategies simply take too long.

FIGURE 6.4—(*continued*)

Teachers' Image of the Whole-Class Discussion

During the discussion, introduce a way to use the solutions to draw the 100th pattern, for example:

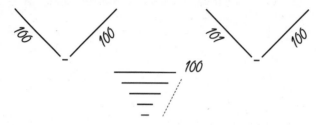

Formula Discussion (10 minutes)

Ask students to take about three minutes to find a formula if they weren't able to prior to the discussion. Tell them to use P for the pattern number and T for the total birds. During those three minutes, have some students place a variety of formulas on the board. Possibilities:

$2P + 1 = T$ (**Solution 1**)
$T = 2P + 1$ (**Solution 2**)
$T = (P + 1) + P$ (**Solution 3**)
Next = Now + 2 (**Solution 6**)

Ask: "Which solutions are correct?" (Have students use smiley faces again.)

Topics that might come up:

• The equivalence of equations (they all arrive at the same amount but are written different ways); $T = 2P + 1$ is actually from two different pictures. In Solution 1, the 2 stands for 2 arms, and the P is the number of dots in each arm. In Solution 2, the P stands for how many sets of 2 are in the tornado (or Wi-Fi symbol).

• In the third equation, $P + 1$ stands for the number of dots in the left arm of the V, and P represents the number of dots in the right arm.

• The final solution is a recursive formula, but it is hard to use in this case with such a big number.

Also, ask students if it matters which side of the equation the T is on. Often students are uncomfortable with NOT starting with the single variable on the left side.

Summarize

Close the discussion by having students write down the big ideas they learned:

1. We can use smaller bird patterns (such as patterns 1, 2, 3, 4, and 5) to predict the number of dots in later patterns that are too large to count.

2. Sometimes the bird pattern might contain a "leader bird" that appears only once in the pattern, but the number of other dots can grow.

3. Labeling the pattern (finding the Pth pattern) might help you write an equation.

4. Many formulas can be used to summarize the bird patterns; they are all equivalent.

5. The Next/Now formulas are called *recursive* (they repeat indefinitely), but they are not very useful for big numbers.

FIGURE 6.5

FIGURE 6.5

Launch of the Birds/Linear Patterns Investigation

Launch (Task presentation)

Engage students by showing a YouTube video called *Wisdom of the Geese (Motivational).* Talk for a few minutes about the meaning of teamwork as shown by the geese in the video. Explain that this video inspired the mathematics problem we are going to work on today. Present four spatial patterns of dots that represent birds flying in "V" shapes. The first pattern contains three birds; the second, five; and so on.

We will ask them to determine the number of birds in different pattern numbers and perhaps write an equation for any pattern number, *P*.

FIGURE 6.6

Formative Assessment for the First Page of the Birds/Linear Investigation

Formative Assessment

Assign as either an exit ticket or homework:

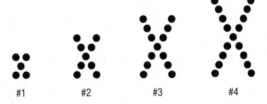

#1   #2   #3   #4

Draw the fifth, sixth, and seventh patterns. How many birds will be in the 10th pattern? Put some evidence on your paper to prove it.

How many birds will be in the 100th pattern? Put some evidence on your paper to prove it.

CHALLENGE: Write an equation to show the relationship between the pattern number and the number of birds.

When Lesson Images Differ

As you worked through this exercise, your lesson image probably did not match Ms. Cline and Ms. Allred's perfectly. Keep in mind that they have had a couple of years to gain experience in lesson imaging, so their image might be more detailed and diverse than yours. Also, there is nothing to say that their image is the "correct" one. After all, when planning a 50th wedding anniversary, it's unlikely that one person's image of the perfect celebration will look the same as another's. That doesn't make one of the images wrong;

it simply means that people imagine events differently and generally prefer to carry them out in their own way. The important point is that a party, or a lesson, is more successful if it is imaged beforehand; we are more likely to work toward supporting our goals than if we had not imaged it at all.

In fact, Ms. Cline and Ms. Allred have co-imaged lessons like this one in a team of 7th grade teachers, and even though the team agrees on a shared lesson image, it always turns out differently—both across teachers and within a teacher's own classrooms, if he or she teaches more than one class period. Each teacher implements lessons with his or her own style and understanding of the goals, and each class comprises many different personalities—Ms. Cline's first period class might enact the image differently from her third period class. (As a matter of fact, Ms. Cline reports that her first class did not come up with the Wi-Fi method, but her other class did. Thus, she was able to capitalize on that strategy in one class but not in the other.)

Classroom Enactment of the Lesson Image

The teachers launched the task with the geese video and asked students to think about how geese culture relates to their work with one another in the classroom. One student remarked that the geese all worked together as a group, and that's what Ms. Cline wanted them to do. Another student fixated on the role of the leader and how that changed throughout the flight. She stated that all the students need to take turns being a leader in class. Another student said he liked how some of the geese would fly back to help a tired bird and help him get back to the rest of the flock. He took that to mean that students should help out one another in class. Ms. Cline agreed, saying, "We are only as strong as our weakest link. It's not about competition in here—it's about helping everyone along the flight."

Ms. Cline then launched the bird lesson:

Ms. Cline: So, I just happen to live—it's not in the country, but back behind our house is Rocky River, and I like to sit out there when it's nice weather like this. And every once in a while there are a lot of geese that come to the bottom down near the river to fish or do whatever they do, and I notice them flying in this V pattern. Has anybody else ever seen them fly in a pattern similar to this?

About seven hands go up.

Ms. Cline: A couple of you? All right, so that made me think, "I could use that as a math problem to get my kids thinking about algebra." So, this is how I visualize the birds. [She points to the V pattern of dots on the board.] If I call those dots birds, that's why. You've got four different pictures up at the top of your sheet. I want you to just take a second and look at those pictures and then answer the first three questions on your sheet. So, everybody read those real quick and see if you have any direction questions. Everybody understand your jobs? Draw the next three for number 1. How many birds would be in pattern number 10? And tell me. And then how many birds would be in pattern number 100, and tell me. First couple of minutes, I want you working on your own. And then I'll tell you when you can start helping each other a little bit.

Take a minute to study Ms. Cline's launch. What do you notice that is similar to and different from her original lesson image?

From our point of view, she altered the launch a bit by relating it to her own experience rather than just the video. She asked students if they had ever seen birds fly in a pattern like this to help them draw on their own personal experiences to relate to the problem. This relates to the key criteria for launching tasks: ensuring that students understand the context. Ms. Cline also let the students know what the picture represents: dots are birds. She made sure that students did not have questions about the task and then had students work independently before touching base with their groupmates.

Regarding that last technique, teachers may choose different organizational strategies for small-group work. Ms. Cline clearly thought that it was important to give individual students a few minutes to process the problem before collaborating with someone else, so they would have a better chance of engaging more fully in the group work.

During explore time, Ms. Cline and Ms. Allred walked around the room, spending no more than two or three minutes per group. Their intention was to collect data on the ways that students were structuring the bird patterns to find answers to subsequent pattern numbers. We present one of the monitoring sessions below, so you can analyze the interactions between the teacher and students Cameron and Selena:

> **Ms. Cline:** So how are you thinking about this one?
>
> **Selena:** I just added two. I saw the pattern was adding two.
>
> **Ms. Cline:** Oh. So you saw 7 + 2 is 9, and that's what these numbers refer to?
>
> **Selena:** Yes.
>
> **Ms. Cline:** That's a great strategy! So what are you gonna do down here at 100?
>
> **Selena:** I don't know.
>
> **Ms. Cline** [to Cameron]: Do you?
>
> **Cameron:** I was gonna say, like, whatever you get for the 10th problem, why don't you just multiply that by 10?
>
> **Ms. Cline:** All right, why don't y'all try that?

Note that Ms. Cline began this session with an answer-unknown question: "How are you thinking about this one?" Ms. Cline does not know the answer and is genuinely interested in the student's thinking, not just her answer. Selena shares that she continually added two until she finished drawing her patterns, which is one of the solution strategies that Ms. Cline and Ms. Allred had anticipated. Ms. Cline then asked a follow-up question to determine if Selena had a way to either use her strategy or adjust it for a larger pattern number, 100. That's when her partner, Cameron, interjected with an idea to take the number of birds in the 10th pattern and multiply it by 10. Again, this was a strategy anticipated by the teachers. As we know, this strategy will return an incorrect answer, but instead of asking funneling questions to get the students to see the error in their strategy, Ms. Cline encouraged them to continue along that path.

Many teachers may not be comfortable leaving students with an incorrect answer or strategy, but Ms. Cline knew that this method would come up in the whole-class discussion, where it could be debated, perhaps by

Cameron, and corrected. Additionally, she knew that students had recently worked on ratios in which doubling and tripling were legitimate strategies because there was a zero y-intercept involved (e.g., if 2 boys can eat 5 hot dogs, then double the boys [4] can eat double the hot dogs [10]). Discussing this strategy in class would give students a chance to revise their thinking about ratios having a special linear relationship, one with a zero y-intercept.

Ms. Cline's next group included three students who had different ways of participating:

Rosa: You multiply the pattern number by two, and then you add one.

Ms. Cline: Why do you do that?

Rosa: Because the number of the pattern, there's always that many right here [traces each wing]. And then there's always one extra right there [points to the leader bird].

Ms. Cline: Cathy, did you think about it the same way she did?

Cathy: No. I did it way different.

Ms. Cline: OK, so talk to me about what you did.

Cathy: It depends on the pattern number. You would, like, go up one. So if the pattern number was five, you would do, like, 6 + 5. You wouldn't double-count it [referring to the leader bird].

Ms. Cline: Could you show me that on one of the pictures if I asked you to? What do you mean?

Cathy: Like, on three, you would do 1, 2, 3, 4 + 3, because you don't recount the bottom one.

Ms. Cline: I like how you said that. Now, what about you, Justine?

Justine: I didn't really get it.

Ms. Cline: Well, can you talk to your groupmates?

Justine: Yeah.

Several teacher moves are significant to point out here. Ms. Cline's first question to Rosa focused not just on the calculations Rosa did but also on *why* she did them. When Rosa says she multiplied two times the pattern number, Ms. Cline could have been satisfied with that correct answer. However, she pushed Rosa to explain where those calculations came from, which Rosa did by referring to the picture of the birds. Ms. Cline had a similar interaction with Cathy and even asked Cathy to explain her strategy using a picture. (Ms. Cline then made a note on her tablet that she had found students using two of the strategies she and Ms. Allred had anticipated.) When Justine either did not have a way to engage in the problem or had chosen not to do so, Ms. Cline did not stay with Justine until she'd figured it out; instead, Ms. Cline trusted Rosa and Cathy to work with Justine to find a way to tackle the problem, encouraging group work.

Ms. Allred, for her part, also walked around the room, monitoring and taking notes. Before Ms. Cline began the whole-class discussion, the teachers huddled in the front of the classroom to compare notes and decide whom to call on for each strategy. They decided not to discuss the pictures for the next four patterns but, rather, to have students share their ways of determining the number of birds in the 10th pattern. They believed that starting with the 10th pattern would probably compel students to discuss their drawings anyway, so they could save time by starting with problem 2. They decided that Ms. Cline would lead this discussion, and Ms. Allred would lead the next whole-class discussion.

Ms. Cline: How many people got 21?
About half the students raise their hands.
Ms. Cline [to a student whose hand isn't raised]: Did you get 21?
Student: No.
Ms. Cline: No? Jada, did you?
Jada: No.
Ms. Cline: We had some really interesting strategies on that one, and then some of you had to abandon those. So, Ethan and Nancy, the two of you kind of thought about it along the same lines. Can you two tell us what you were thinking to get the 10th pattern? Or what you noticed was happening?
Nancy: You add 2 every time you draw one pattern.
Ms. Cline: OK, and what I saw on her paper, she had a 3 here and a 5 here. She had a 7 here, and she had a 9 here. (See Figure 6.7.)

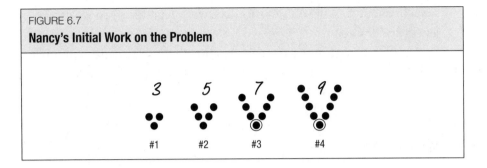

FIGURE 6.7
Nancy's Initial Work on the Problem

Ms. Cline: All right, Ethan, he doesn't want to brag, but he even made a chart. Now, some of you also had some great strategies. Jenel, can you talk to us about what you were picturing, and you can even come draw on my birds if you think that'll help. How did you see the pattern?

Jenel: I saw the 1 in the middle, and then for each one it was the pattern number, so in this one [points to pattern 3], it's like the pattern number times 2.

Marilyn: That's what I did!

Other students echo: That's what I did!

Marilyn: I did P × 2 + 1.

Ms. Cline: Ah! Does Jenel's way make a little bit of sense?

The students affirm that it does.

Ms. Cline: I call this a certain method—and if this is the flock of birds, what is this guy's job?

Various students: Leader.

Ms. Cline: He's the leader, so I call this the Leader Bird method. Who did the Leader Bird method? Raise your hand high if you did the Leader Bird. You pictured that little guy, and then the pattern number was the other two [motions hands to show the wings coming out from the leader bird]. All right, so look at all the leader birds.

About six students raise their hands.

Ms. Cline: OK, that's pretty good. There were a couple of other ways to think about it. Now, I want Cathy to come up and show them how you thought about it. Will you show us your way on pattern 4 so we can keep Jenel's leader bird?

Cathy: So I did one above the pattern number plus the pattern number. So I did—because it was pattern number 4, I did 5 + 4 [Figure 6.8].

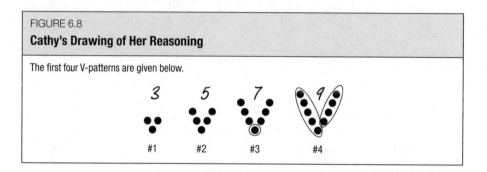

FIGURE 6.8
Cathy's Drawing of Her Reasoning

The first four V-patterns are given below.

Ms. Cline: So, what's the relationship between the two things that you circled?

Cathy: Like, you can't count the bottom again.

Ms. Cline: OK, did you guys hear what she said? What did she say?

Melissa: You can't count the bottom again.

Ms. Cline: Yeah, you're not going to count that leader bird twice, are you? So which one of Cathy's ovals, which group, the group of four or the group of five, is indicated with the pattern number? Which one, Staci?

Staci: Four.

Ms. Cline: The one with four. And how would you describe the one with five?

Tom: The pattern plus one.

Ms. Cline: The pattern plus one. So, we've got the Leader Bird, and I call Cathy's the Plus One. Ah, and I don't think I got anything different.

As in the last excerpt, there are several notable moves made by the teacher here. Ms. Cline begins the conversation by asking what the answer is, and luckily a couple of students disagree with the answer. This gives Ms. Cline the chance to have students justify their answer of 21. Notice that Ms. Cline does not ask just any students to volunteer their thinking but, rather, calls on Nancy and Ethan directly. She knew that they had started with the strategy most common to this class, recursive reasoning. Knowing that the 10th pattern is small enough to make a chart like Ethan's without any difficulty, she moved quickly to the next strategy, Jenel's, the Leader Bird method. This strategy obtains the same answer but uses a certain visual structuring that might be more efficient than a chart. The efficiency and usefulness of the strategies will be discussed later, once all the strategies have been presented.

Notice also that Ms. Cline names the strategies, which makes them easier to remember in future problem solving. We have found that naming strategies, either with the student's name (Jenel's way) or with a more explicit name (Leader Bird), is a powerful way to get students to participate as well as to recall prior strategies.

Ms. Cline then calls on Cathy, whom she knew from monitoring had the Plus One strategy. All students used the drawings to make a record of their thinking more explicit for other students. In order to bring the level of the discussion higher and to address the inefficiency of the recursive method, Ms. Cline asks the following question:

> **Ms. Cline:** How many of you had to change strategies from question 2 to question 3? I asked you how many birds were in the 10th pattern and then how many were in the 100th. Did anybody have to change their strategy? Sam says she had to change her strategy. All right, if you tried pattern plus two and I asked you to get to 100, what was that like for you?
>
> **Student:** It would take a while, a lot of paper.
>
> **Ms. Cline:** It would take a while, it would take a whole bunch of your paper? What if I said to you, now that you've seen some of the strategies, I want you to talk to the people at your table, and over to the side I want everybody to figure out how many birds are in the 1,000th pattern. Go!

In this bit of discussion, the teacher said a lot of words, but she didn't tell students which solution methods they should use. Instead, a student mentioned the difficulty of using the recursive method for pattern 100, and the teacher challenged students to use a method they had discussed to find the number of birds in the 1,000th pattern. As students solved this problem, Ms. Allred walked over to Ethan's group to determine how he was reasoning.

> **Ethan:** I don't even know what I'm doing!
>
> **Ms. Allred:** OK, so whose way did you understand the best?
>
> **Ethan:** I don't even know—it's confusing!
>
> **Ms. Allred:** OK, so what's confusing? Ask a good question.
>
> **Ethan:** Like the Leader Bird, I don't even know what she did. Do you, like, 3×2 if for, like, for pattern 3, would you do, like, 2×3, because there are two rows of three?
>
> **Ms. Allred:** Yeah! Because what you were doing earlier, you were trying to count that leader bird twice, and she said you can't count that guy twice. You have two ones here [in pattern 1]. What do you have here [pointing to the second pattern]?
>
> **Ethan:** Two twos. Two threes… two fours.
>
> **Ms. Allred:** OK, so use—if that makes sense to you, what would you do to find 1,000?

Both teachers found students who were still processing the strategies they had heard during the whole-class discussion. Rather than reteach the methods, Ms. Allred attempted to meet Ethan at his level by asking him which strategy made the best sense to him. When Ethan said he did not know, Ms. Allred could have easily given in and retaught the strategies. Instead, she asked him to find the part that was confusing and then ask a good question about it. This teaching technique puts the onus of learning back on the student and teaches him how to ask questions when he does not understand. He learns that he is the source of his learning, not the teacher, and that he needs to learn how to ask questions to further his understanding.

Ms. Cline made a quick decision that the class should discuss the 100th pattern before concentrating on the 1,000th. Before many students had finished, she called them back for a whole-class discussion. She began by asking what answers students got, and most students responded with "201."

Ms. Cline: 201? Darren, could you go to the whiteboard, and could you just, real quick, show us how you could use the Leader Bird method to get 100, to get the 100th pattern? What would it look like?

Darren: You could do, like, 100 on the stem. Then, like, 200, and then plus, add the leader bird of course, and then 201 [writes "100 + 100 + 1" vertically].

Ms. Cline: All right, so Darren's got his math brain rolling. He doesn't even need a picture. Can somebody hop up there and show us what it would look like if you were to sketch it? To help you figure it out? Nobody?

Pam: I can try.

Ms. Cline: Go try! That's what I want to see!

In this part of the discussion, Ms. Cline appreciated what Darren put on the board; however, she knew that other students might not connect with his calculation on its own and that a visual representation might be helpful for them. Because no student wanted to draw 201 dots, Pam offered to try to create a viable picture. This episode underscores the importance we place on imaging the representations or symbols that either the teacher or the students will use during problem solving. Ms. Cline knew that visual representations of the different ways in which students were structuring the V pattern were paramount to making sense of the various strategies. Numbers alone, like Darren's method, would make sense only to those who already understood his structuring.

Pam draws the picture in Figure 6.9.

Ms. Cline: Anybody else picking up on Pam's method? No?

Shane: It's like you can't really get an exact number. It's hard.

FIGURE 6.9

Pam's Drawing for the 100th Pattern

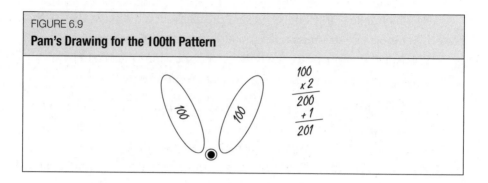

Ms. Cline: Ah, you think it's hard. Cathy, could you come over and show us your picture, what your sketch would look like, to get 100? Pam, stay up there.

Cathy draws the picture in Figure 6.10.

FIGURE 6.10

Cathy's Drawing for the 100th Pattern

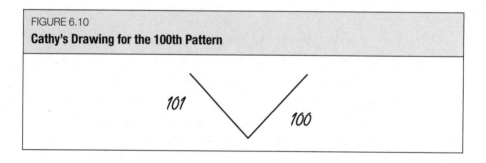

Ms. Cline: OK, so, Pam, talk to us about what you put up there. And guys, this is really, really important if you want to be able to move forward, so listen to Pam and be ready to ask her a question if you need to ask one.

Pam: I don't know what to say. I just drew a V, and I put 100 on one side and 100 on the other, and the 1 on the end [points to the leader bird].

Ms. Cline: And who does that one represent?

Pam: The leader bird.

Ms. Cline: OK, so that's the one leader bird. Any questions for Pam? What do you think about that? Is that good, you're not sure? Or you have no idea? Let me see who's got it—yes, I got it, thumbs up; thumb down, I don't think it's right; or to the side. All right, I see a lot of ups and a few to the side. So, if you're to the side, let's look at Cathy's and see if hers makes any more sense.

Again, note the questions that Ms. Cline asked to help students who did not understand the Leader Bird method to make sense of it during this conversation. Not only did Pam draw a reasonable representation of the Leader Bird method, but the teacher asked what the lone dot stood for in the context of the problem. The teacher then moved to Cathy's drawing for the Plus One method:

Cathy: So you do the pattern number up, which would be 101 + 100 because you can't count the bottom one twice.

Ms. Cline: Are you looking at it like she is? Did you also get 201? OK, does anyone have a question for Cathy?

John: What if, like, the number is bigger than 100? It's like something like 1,000?

Cathy [and another student, speaking concurrently]: Then you do 1,001 + 1,000.

John: Oh. Never mind.

Ms. Cline: Do you know why she's doing 100 and then 101?

John: Because you can't count the leader bird twice.

Ms. Cline: Yeah, you can't count that little guy at the bottom twice, can you?

Cathy draws Figure 6.11 on the board while Ms. Cline talks.

Ms. Cline: I saw where—Darren, was it you? No, it wasn't. It was somebody that did this. So, John, when we make that line… [Ms. Cline draws Figure 6.12].

Ms. Cline: This is our little—one extra. So, in pattern 2, there were how many there? Two and two. In pattern 3, there were three and three, in pattern 4, there were four and four, until we get to that leader bird. So does that help you make sense of Pam's? Now, I want you to go back and finish that 1,000th. If you've already finished 1,000, I want you to figure out pattern 150.

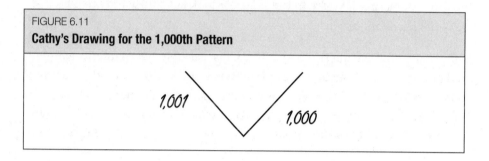

FIGURE 6.11
Cathy's Drawing for the 1,000th Pattern

Here, Ms. Cline used the pictures of the bird patterns on the board one last time to reiterate the way that Pam was structuring the V patterns. Importantly, the teacher drew a line that separated the leader bird from the two wings and restated that the wings have the pattern number in them: "2 and 2," "3 and 3," and so on. She added the line for extra visual support, which might later lead to understanding a linear equation as structured by a constant starting amount and changing amount.

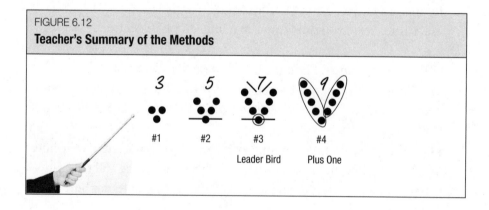

FIGURE 6.12
Teacher's Summary of the Methods

At this point, 26 minutes into the class period, the conversation turned back to finding the number of birds in the 1,000th pattern, with the hope that more students would be successful after having discussed the 100th pattern. Indeed, most students had a way to quickly find the answer to this question, so Ms. Allred challenged the class to find a formula. To start the whole-class discussion, she asked several students to put their formulas on the board for inspection by the class. The following formulas were placed on the board by individual students, with their name by each:

$$P + (P + 1)$$
$$P \times 2 + 1 = T$$
$$P \times P + 1 = T$$
$$P \times 2 + 1$$
$$(P \times 2) + 1$$

Ms. Allred began by asking, "Can you use one of these patterns [referring to Pam's and Cathy's methods, which had been left on the board] to explain your formula?" She then asked to students to work in their groups to determine which formulas were viable and which were inaccurate.

For the rest of the 20-minute session, Ms. Allred led the whole-class discussion by having students indicate agreement or disagreement with each formula. Students compared and contrasted the formulas to determine that the first formula is the same as (equivalent to) the second, fourth, and fifth formulas. They noted the distinction between formulas that had $= T$ and those that did not have an equal sign; while both represent the same quantity, the teachers noted that the former is an equation, and the latter is an expression. In this way, equivalent expressions and equations and the difference between equations and expressions were discussed, which were subgoals of the main objective of the day.

To end the class, Ms. Allred assigned the X pattern as an exit slip to determine how individual students were reasoning and to have data to help plan for tomorrow's class. We show a few of the students' solutions to illustrate the diversity of reasoning that still existed after one day of the unit (Figure 6.13).

FIGURE 6.13
Four Different Students' Exit Slips

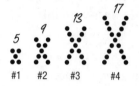

#1   #2   #3   #4

Draw the fifth pattern. How many birds will
be in the 10th pattern? Put some evidence
on your paper to prove it.

How many birds will be in the 100th pattern?
Put some evidence on your paper to prove it.

STUDENT 1

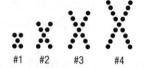

#1   #2   #3   #4

Draw the fifth pattern. How many birds will
be in the 10th pattern? Put some evidence
on your paper to prove it.

How many birds will be in the 100th pattern?
Put some evidence on your paper to prove it.

STUDENT 2

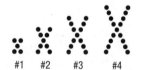

#1   #2   #3   #4

Draw the fifth pattern. How many birds will
be in the 10th pattern? Put some evidence
on your paper to prove it.

How many birds will be in the 100th pattern?
Put some evidence on your paper to prove it.

STUDENT 3

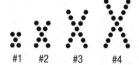

#1   #2   #3   #4

Draw the fifth pattern. How many birds will
be in the 10th pattern? Put some evidence
on your paper to prove it.

How many birds will be in the 100th pattern?
Put some evidence on your paper to prove it.

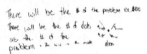

STUDENT 4

After class, Ms. Allred and Ms. Cline analyzed the exit slips:

- Six students' papers (25 percent) resembled Student 1's, in that there was no solution found.
- Of the remaining students, 18 were able to find answers to all four questions, and 12 of those included a formula, such as Student 2's and Student 4's.
- Of the formulas, two were equations and 10 were expressions.

It wasn't clear to the teachers whether those who did not write formulas had run out of class time or were unable to do the problem. Because 75 percent of the class was able to solve the problem for large numbers, the teachers decided to continue with more bird patterns the next day. They would begin with a simple Y pattern and then move to patterns where the leader bird included more than one bird.

Conclusion

In this chapter, we provided an opportunity for you to lesson image one class period that had already been imaged by two experienced teachers. We noted that no two lesson images have (or are likely) to be the same because teachers are unique and have different ways of interpreting and implementing lessons. We presented each component of the lesson image and included snippets of actual classroom dialogue from the teachers' implementation of their lesson image. We stopped along the way to point out aspects of the lesson image that were followed as written and aspects that were altered in action. The main point is that no lesson image flows exactly the way we expect it to. Nonetheless, lesson imaging better prepares inquiry teachers to interpret students' reasoning and to know what questions to ask in the moment of teaching.

Before Reading Chapter 7...

Consider these questions before moving on to the next chapter:

• What are some steps you need to take in order to incorporate lesson imaging into your practice?

• What constrains you at the moment from making such a shift?

• What supports are necessary for you to lesson image with your peers?

• What role do the principal and coach play in lesson imaging?

7

Getting Started

Lesson imaging is different from what many teachers do when they plan their lessons. Lesson imaging requires effective collaboration among a group of teachers in an environment of trust. Teachers must be willing participants in the process; the process assumes that teachers are engaged in growing professionally on an ongoing basis, using student learning to identify problems of practice they wish to address and improve, so they can provide improved opportunities to learn for the students.

—Robin Dehlinger, principal, Florida

Transforming one's lesson planning sessions into lesson imaging takes knowledge, practice, motivation, commitment, and administrative support. While not impossible, it is very difficult for STEM teachers to lesson image by themselves without the support of administrators and without a coach or mentor. As Cobb and McClain (2006) posit, a teacher's practices and decision making are situated within various other groups. Teacher change is enabled and constrained as teachers form networks that are embedded within the confines of other groups. In other words, teachers who want to shift their practice toward teaching for autonomy are both constrained and enabled by their relationships with others within the school, as well as by the types of support that are available.

As Figure 7.1 illustrates, a single STEM teacher does not actually work in isolation. He or she must work across boundaries, interacting in meaningful ways with others in the department, with other teachers at the school, and with the administration and other staff members. Those interactions enable and constrain the teacher's teaching practices, whether the teacher is cognizant of it or not.

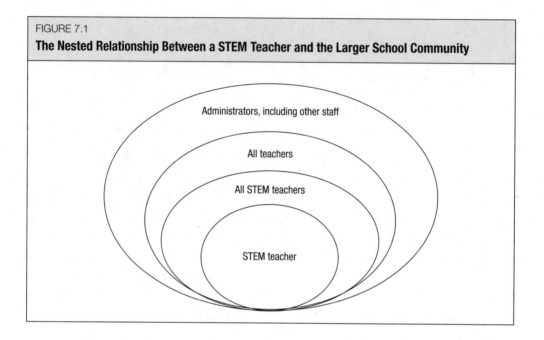

FIGURE 7.1

The Nested Relationship Between a STEM Teacher and the Larger School Community

For example, a science teacher, Mr. Moran, may read this book and decide that he wants to begin teaching science for autonomy and to implement lesson imaging in his planning practice. He shares this with his fellow science teachers, and two of them get excited as well. However, Mr. Moran has seventh period planning, and the other teachers have second and third period planning. Teaching schedules have therefore constrained their potential collaboration and pedagogical shift. It will take meeting with the assistant principal in charge of scheduling to convince her to create a common planning period in the next school year to enable them to co-image.

In this chapter, we discuss the supports needed to establish an educational setting that is conducive to teaching for autonomy in general and to lesson imaging specifically. We elaborate on the challenges we have come up against in attempting to make schoolwide shifts toward this type of teaching and planning. We explore all four levels depicted in Figure 7.1, beginning with what it takes for an individual to get started. We consider the challenges associated with establishing small groups of teachers who co-image on an ongoing basis. We argue for the importance of STEM coaches to support teachers in their difficult shifts in practice. Finally, we outline the role of administrators and staff in supporting, and sometimes constraining, the efforts of teachers to lesson image. Throughout, we share our experiences with supporting lesson imaging among colleagues across two different states and a dozen different schools.

Teacher Change

What are the inspirations—the catalysts—that provoke teachers to make changes in their practice, especially when their students' test scores are high? Even if test scores are less than desirable, what are the characteristics of teachers who are comfortable with change? Whether ready for change or not, what is the best way to begin the process, especially when working with an entire STEM faculty?

In this section, we focus on the different types of teachers we have worked with, from the most gung ho to the most resistant. Included throughout are stories from the authors, who are working within schools to promote and support changing teachers' practices toward teaching for autonomy and lesson imaging.

The Reluctant Teacher
Chris Cline

One reason that teachers change how they do things is that a change is mandated by the administrative team. In our building, our principal wanted to see an increase in students' *level of engagement* in the mathematics classroom. She also wanted to see students at the board more and to hear students constructing and defending their own solutions. Our principal decided to funnel resources and professional development time toward these strategies. When big decisions like this are made by principals, they can be very unpopular. If this is the case in your building, like mine, expect to hear, "Eighty percent of my students are proficient on the end-of-year assessment," or, "Why do I need to change? I'm a good teacher!" Both of these responses are legitimate and express exactly how I felt.

In our building, the entire math department came together in the summer of 2013 to discuss a different way to deliver the mathematics content to our students. With guidance from a professor at a local university, our principal had a vision of moving from the traditional classroom that was controlled by the teacher to a more student-centered classroom.

I was not happy; I took this change to mean that I was not doing my job as a teacher, and I knew that I was one of the best. I met this philosophical change with much skepticism. Because this delivery of instruction was totally different from mine, I did not want this change to be effective—in my mind, if this worked, then I really was not the best and should have been doing a better job. I took it all very personally. I did not change how I taught my classes for the first couple of months during the fall of 2014.

However, as a group, the 7th grade math teachers decided that we would attempt to teach our ratios and proportions unit for autonomy. With the help of the university professor, we began meeting to image this unit. We talked about the many different ways our students could and would answer these questions. We met once a week for a couple of weeks to explore and discuss the different student solutions. All the while we were imaging, I kept hoping that this method would flop and that my kids would not understand the concepts as completely as they would have if I had been in complete control of the classroom. However, I knew that in order for me to say that my students did not understand because of the new method of delivering the instruction, I had to do my best in staying with the plan developed in our math meetings.

Mr. Cline's reluctance to change his practice is not uncommon among classroom teachers. He readily admits that he was offended by the suggestion that his current practice (traditional instruction) was insufficient for student success. In fact, Mr. Cline's algebra students were consistently at the top of his district in terms of test scores, so one can imagine his confusion when he learned that the mathematics department was going to try a new pedagogical approach. Luckily, Mr. Cline determined that if the new approach was going to fail, it would not be because he tried to sabotage it. He would implement the approach fully, and if it was not successful, it was not because he tried to make it fail—it would fail because it was not as good as his previous teaching. With the support of his 7th grade colleagues and the university professor, he engaged in multiple meetings in which the teachers lesson imaged with *Mathematics in Context* (Romberg & de Lange, 1998). Mr. Cline continues his story.

My Students Were My Catalyst
Chris Cline

After using the imaging from our meetings, I became a convert in about three class periods. I had been teaching for almost 20 years and had always gotten really good results from end-of-grade tests; however, the discussions that my students were now having about mathematics were unreal. I had never really had students talking and on task as much as they were using this new approach. My students were always engaged, but this was *way* different. The main reason for wanting to completely overhaul my method of delivering instruction came down to the students. Witnessing the level of their engagement when teaching for autonomy was fully implemented was amazing. Because of the imaging with the 7th grade math teachers, I was able to predict *many* different student responses, not just the one or two that I would generally show them. With this method, students were sharing their explanations, and they credited their peers by naming the solution process after the student, such as "James's method."

Simply listening to his students won Mr. Cline over in just three class periods. Hearing his students engage in a different type of mathematics talk inspired him to commit more fully to lesson imaging and teaching for autonomy.

One of the first steps in supporting teacher change is to help teachers develop an ear for listening (Stephan, Underwood-Gregg, & Yackel, 2014). *Listening interpretively* means engaging students in a dialogue in which you hear how they are reasoning about a solution. *Hermeneutic listening* occurs when teachers listen interpretively and in such a way that their own mathematical knowledge changes (Davis, 1997). We have found that when teachers listen more interpretively and hermeneutically, rather than listening *evaluatively* to assess whether a student's answer is correct, they may see a need to shift their practice toward autonomy. Mr. Cline began listening interpretively to his students, as can be seen by both his reaction to the sophistication of his students' methods and the importance he placed on anticipating students' reasoning that might be different from his own.

The Importance of Lesson Imaging
Chris Cline

As kids began to form learning teams, their level of engagement increased exponentially. They were able to have conversations in these teams about their different solutions, and they were able to critique the solutions of others in their group. As they became more at ease in sharing their responses, they would say things such as, "I don't think I'm right, but this is what I think...". I could see them becoming more interested in math while at the same time becoming better mathematicians. I was more a facilitator of the learning that was taking place within my classroom and less a dictator of how learning should be taking place.

When I stepped back and evaluated what was transpiring within my classroom, I was in awe of the students' ability to have conversations about math, their ability to reason, their ability to problem solve—and they were doing all this without me telling them how to do it. They were in control of their own level of understanding. I saw students who were drawing pictures to answer questions such as, "If 2 aliens eat 7 candy bars, how many candy bars can 10 aliens eat?"—but along with those drawings that we displayed on the board, other students were able to also display more abstract algebraic proportions. In seeing all the different solutions to this problem, students began to see the connection from their particular method of solving the problem to the more abstract method.

One of my biggest problems was neglecting the more concrete approaches and just starting with the abstract—or I would just focus on the concrete approach, because the students in that particular class would not be able to understand the more abstract approach. In doing this, I always failed to address a group of students within the classroom. It was through the imaging process that I was aware of all the different approaches to look for and the connectedness of these approaches. In using this approach, I no longer neglected one group of students but was able to finally meet the needs of *all* my students.

Not all reluctant teachers turn around quite as quickly as Mr. Cline did, and some never do at all. Author Michelle Stephan has worked with some teachers who still do not see a need to teach for autonomy after 10 years of exposure and multiple levels of support. We discuss how to handle these teachers in a subsequent section about administrators' roles in teaching for autonomy.

While researchers have not determined the best possible learning route for supporting teachers on their journey to teach for autonomy, Akyuz, Stephan, and Dixon (2013) have written about the planning and teaching practices that are paramount to this approach (Figure 7.2).

Daily lesson imaging and long-term visions of the content to be taught are paramount to successful teaching for autonomy. Formative and summative assessments are the vehicles for collecting data to inform subsequent lesson imaging. Stephan, Underwood-Gregg, and Yackel (2014) outline a possible learning route for teachers and suggest that teaching teachers how to listen to their students may serve as a first learning goal as well as an inspiration to change. Becoming familiar with the research on how students learn a concept, the planning practices, the classroom practices, and coaching techniques involves subsequent learning goals that can support full implementation of this teaching approach. Because the main point of this book is to learn to lesson image, we leave it to readers to investigate these learning goals.

FIGURE 7.2

Planning Practices for Teaching for Autonomy

1. *Prepare:* Create a long-term image of the content to be taught in the unit, which can be done by reading research on how students understand the content, working through the tasks/activities/labs in the unit, and unpacking the standards.
2. *Look forward:* On a daily basis, lesson image each class period to be taught in the unit.
3. *Look backward:* After teaching the lesson, reflect on the students' reasoning, and consider changes that should be made to instruction.
4. *Assess:* Analyze formative and summative assessments.
5. *Revise:* Alter lesson images and materials, both in real time and after the unit has been completed.

Source: Adapted from "Improving the Quality of Mathematics Teaching with Effective Planning Practices," by D. Akyuz, M. Stephan, and J. K. Dixon, 2013, *Teacher Development*, 17(1), pp. 92–106. Copyright 2013 by the Taylor & Francis Group.

Communities of Learners

What roles do people at the other levels of Figure 7.1 play in creating an environment conducive to lesson imaging and teaching for autonomy? In his groundbreaking work, DuFour (2004) argues that if school reform is to be sustained in the long term, faculty, staff, and administrators must form what he calls *professional learning communities* (PLCs). PLCs share three key characteristics: they create a shared vision for student growth, commit to collaboration, and use data to improve learning.

As with many compelling educational concepts, the term *PLC* has been appropriated by many teachers and administrators at all levels in many differing ways. We consider it to describe an entire school—principals, counselors, teachers, and other school staff—not small teams of people. In addition, a PLC is more than a team of teachers who meet regularly; the team must also embody the characteristics identified by DuFour: collaboration, shared vision and norms, and data analysis to inform decisions.

We use the term *communities of learners* (COLs) to describe smaller teams within a school that embody the characteristics of a larger PLC (Stephan, Akyuz, McManus, & Smith, 2012). COLs are teams of two or more individuals (teachers, staff, and administrators) who meet at regularly scheduled times to collaborate about students' learning and teaching practices and to analyze data (both quantitative and qualitative) to inform instruction. The COL structure is one of the most important supports for enacting lesson imaging as a major pedagogical practice. Julie Cline reflects on her school's process of developing COLs within the mathematics department.

Choosing the Facilitators
Julie Cline

Establishing PLCs was a districtwide initiative when we began our process of lesson imaging, so we were already accustomed to meeting with one another and co-planning. However, once our principal decided that mathematics teachers should shift to imaging and promoting student discourse, she had to restructure COL time and change how the facilitator role was filled. Initially, teachers either volunteered to be the facilitator or were nominated and then voted on—a system that did not always result in the best leader for the group.

As departmentwide implementation of teaching for autonomy was mandated, COL facilitator roles were assigned by the principal, who gave careful thought to which teachers possessed the leadership qualities necessary for this role. If possible, the facilitator would have experience and/or knowledge of the imaging process and leading math discourse in a classroom. In the early stages, few teachers had any experience with this approach, so a good leader in this case would be open to it and have a vision of the changes that needed to be made in teachers' current practice. Another important factor in selecting the facilitator was teaching experience. Experienced teachers can rely on previous interactions with students and their thinking to help in the anticipating portion of imaging. The facilitator must know the standards and make sure that they are at the forefront of all tasks and activities that are selected. Because many times teachers gravitate toward the activities they have always done, a strong leader has to hold the group members accountable for teaching the standards and for autonomy.

First and foremost, Ms. Cline notes that the strategy for choosing a COL facilitator had to change in order to find the best teachers to lead such a radical shift. As Collins puts it in his book *Good to Great* (2001), great organizations begin by getting the right people on the bus and in the right spot—and the wrong people off. While it is more difficult in education to "get the wrong people off the bus," putting them in different spots is more feasible. For example, in Ms. Cline's school, the principal would no longer let teacher popularity dictate who took the leadership positions. Instead, she considered the characteristics of the personnel she had and then deliberately chose certain teachers to facilitate the newly formed COLs. Important characteristics of potential leaders included imaging experience, teaching experience, openness to learning, and the ability to rein in stray conversations and return the focus to imaging goals. Sometimes, the teacher with the most number of years teaching, or the teacher who has won Teacher of the Year for the school, is not the best fit. In fact, in one school that Dr. Stephan worked in, one of the COL leaders was a relatively new teacher who had proved that he could teach for autonomy and lesson image in his own practice.

Many researchers have written about the characteristics of good leaders, so our intent is not to rehash that list here. We simply stress that principals must know how to choose the teachers within their school who can best lead COLs that intend to teach for autonomy. For our purposes, a good facilitator is a good communicator, knows the lesson image process (and has experience doing it, if possible), teaches for autonomy (or is open to professional development), and can keep COL meetings focused on imaging rather than traditional lesson planning or discussion of specific students.

The Structure of a COL
Julie Cline

We have found it helpful if the COL contains teachers who teach at least one section of the same course. For example, there might be three teachers who each teach at least one class of 6th grade science, so they would form a COL in order to lesson image that particular class. While they might teach other science or mathematics classes as well, it is important to find one class that several teachers share in order to have common conversations. Scheduling common planning periods is critical so that these teachers can meet on a regular basis. Our school mandates that each COL meets once a week for about 70 minutes. This is a protected work time, which means that no meetings or conferences are ever scheduled during this block.

In grades where teachers teach more than one subject, meeting days are set to accommodate them. For example, math and social studies meet on Tuesdays, and English/language arts and science meet on Thursdays. Although 70 minutes sounds like plenty of time to meet, a truly effective COL will need to meet more than this—ideally, daily.

Once the COLs have been formed, the facilitator and the members must work together to establish norms for participation that enable all members to feel safe in analyzing their own teaching practice. Additionally, certain expectations must be determined at the outset of the COL meetings, many of which are simple and are the foundation for most groups:

- Be on time.
- Be prepared.
- Listen to others.
- Be a contributing member.
- Follow the agenda.

One of the most critical norms is that lesson imaging is a priority in all meetings. Given that planning time is limited (typically, 45–80 minutes) and that meetings may occur only once per week, teachers should work on the selected tasks and image several different solutions *before* coming to the meeting. This maximizes the time that can be spent on sharing ideas, selecting, and sequencing.

Another important norm that is sometimes overlooked is mutual trust among members, both socially and intellectually. In a functional COL, members should analyze the data provided by formative and summative assessments. Teachers should trust that their colleagues will be supportive and offer suggestions for improvement, rather than be critical and judgmental. This trust also means that members will squelch criticism and negative talk outside the COL. It is easy to get frustrated when changing one's teaching approaches, and teachers often get caught up in conversations that are critical of those changes. A good facilitator notices when norms are violated both during and outside of meetings and addresses those violations head-on. If repeated attempts at establishing these norms fail, the administration must be brought in to clear the air, establish the expectations for COLs, and, when necessary, make personnel adjustments to alleviate tension and non-normative behavior.

What First Meetings Might Look Like
Julie Cline

In the beginning, most of our COL time involved anticipating student responses, and we became very good at this particular practice. Because our principal had purchased *Mathematics in Context* (Romberg & de Lange, 1998), we did not have to find new units or plan common assessments. Instead, we could focus on creating strong launches, anticipating how students might solve the problems, and engineering what our whole-class discussion would look like. However, we found that we spent almost the entire time working through problems and sharing our ideas—and that was as far as we got. We were missing out on the selecting and sequencing pieces, both of which are crucial to good mathematical discourse. We discovered that if we anticipated student responses prior to the COL meeting, we could share strategies more efficiently and then shift to selecting which responses would add to the development of the concept. Crucial questions we asked during our selecting and sequencing discussions included the following:

- What will we do if we don't get all the responses?
- Which of our anticipated responses can be omitted if students do not come up with them?
- Which responses should we show students if they do not come up with them on their own (i.e., which strategies are crucial to the big idea)?
- Which incorrect responses are worthy of including in the whole-class discussion?

For example, when imaging a ratio and proportion sequence with 7th grade teachers, many of us anticipated that the students would use a ratio table to answer questions about how many food bars would feed a certain number of aliens, based on a given ratio of food bars to aliens. This was the first time this group of teachers had taught this sequence, and after they taught the initial lesson, they were surprised that not a single student had used the table. This led to an impromptu meeting where we further discussed how to proceed because the tables didn't come up. The group acknowledged that the sequence got progressively more difficult, and speculated further where the tables might appear. We also agreed on a place in the sequence where we would show the table if it still hadn't been presented by students.

We have found that when we have to present a strategy that helps develop a concept, we embellish a story about how a student in a previous class came up with this table and then ask what the kids think about the approach. In this way, the students do not interpret our suggestion authoritatively but, rather, attempt to make sense of it in their current mathematical world to decide if it is useful or not.

Ms. Cline's reflections indicate two characteristics that are essential for COLs to be sustainable in the long term: *flexibility* and *comfort with discomfort*. Using the traditional method of instruction, most teachers have become proficient in lesson planning and very rarely have to stray from what was planned. When implementing lesson imaging and focusing on a student-centered classroom, flexibility is a must. It is very common to image a set of lessons and carefully plan how much time each launch and discussion will take. However, the conversation among students may not go quite as planned, and the teacher must adjust accordingly. Perhaps the launch took more time because the students were not as familiar with the context as they needed to be—or, like the example above, students do not invent the strategies that were predicted, so the teacher must adapt in action. Maybe the whole-class discussion took more time than expected with the first period but was shorter in the second period, thereby getting class periods misaligned in terms of time.

With student understanding at the forefront, teachers must become comfortable with the fact that sometimes they may not get to all the problems planned. Conversely, lessons may also go faster than anticipated, and teachers must be prepared for that as well. The more teachers image and have this discourse, the better they get at planning the appropriate amount of time for each lesson. Nevertheless, flexibility in teaching that allows for unanticipated (but productive) discussions to emerge is necessary.

Flexibility
Julie Cline

Teachers must allow for flexibility in planning time outside the protected/scheduled time. Some units may require a COL to meet more frequently than normal due to the nature of the content or the level of understanding for a group of students. A teacher may have encountered a unique solution that was not anticipated, so sharing that in an impromptu COL meeting could benefit the entire group. This happens often with my colleagues and usually occurs after school or in the copy room! I've also had lessons that simply did not result in the strategies we anticipated during imaging; having the flexibility to talk to the COL at a moment's notice gives me insight into how the lesson went for others and helps me prepare for the next lesson.

In most productive COLs, the teachers who form the team learn to be comfortable with discomfort. For many teachers, this process is unlike any method of teaching they have ever experienced, and that alone is uncomfortable. Most of us are creatures of habit, and implementing change can make us uncomfortable. We question why we are changing and wonder if we will be as effective if we do change—what will happen to our students' test scores? There are multiple other sources for discomfort, and a good COL leader is someone who can manage the norms by addressing discomfort in a meaningful and sensitive way.

Sources of Discomfort
Julie Cline

Some teachers' discomfort comes from the difficulty of finding multiple strategies to solve a problem. They find it difficult to step away from the traditional algorithms with which most of us were taught. It could be that a teacher tends to think more visually and has difficulty anticipating strategies that are more algorithmic. Some teachers are also intimidated by the thought of explaining their thinking to peers, especially when there is a group member who has a strong mathematical background.

Other teachers' discomfort comes from giving up the reins during class and letting students lead the problem solving and discussion. Many of us became teachers because we were good at something that most of the country is not, and we want to explain it to students so that they can become better, too. That has traditionally involved finding clever and easy ways to explain concepts to students, rather than finding creative tasks and units to help the students solve problems themselves and explain their solutions to their peers.

Another source of discomfort in the classroom comes from the fear of not being able to understand a student's solution method. What do I do if I am stumped by a student and I look stupid? Or what if that happens when my assistant principal spontaneously observes me? What if my administrator visits on a day where the classroom discussion looks nothing like what I imaged and it looks like I do not know what I am doing? The fear of looking inadequate, especially in front of peers and bosses, is a major source of discomfort that must be managed by the COL and administration as a team.

In summary, COLs do not exist because teachers are mandated to meet at a specific time with a certain group of colleagues. While that might be a necessary step, COLs come to be when a group of teachers create shared goals for student learning and positive norms for participating with one another in some rather difficult work. COLs use student data, both quantitative (test scores, quizzes, formal assessments) and qualitative (daily formative assessments, small-group and whole-class dialogue), to inform instruction. When teaching for autonomy, COLs lesson image together as often as possible (more than weekly is optimal) and have a leader who maintains a focus on imaging conversations. The leader has experience with and openness to imaging and teaching for autonomy and exhibits productive communication strategies while managing the inevitable discomfort that comes with change initiatives.

Mentoring and Coaching

A third area for supporting lesson imaging and teaching for autonomy comes in the form of a human resource. It is well accepted that a variety of mentoring approaches—including coaching, co-teaching, co-imaging, model teaching, lesson study, and professional development programs—can have a positive effect on teacher change. School-based coaching, particularly at the elementary and middle school levels, has become routine professional development. More and more schools are hiring part- and full-time coaches or facilitators to conduct a number of activities, including (but not always limited to) working with teachers in their classrooms. While hiring coaches and facilitators is expensive, many administrators and researchers report a positive impact on student achievement, and research has empirically linked the presence of mathematics coaches to increased professional development opportunities for teachers (Campbell & Malkus, 2010; Costa & Garmston, 1994) and to modest gains in student learning outcomes (Campbell & Malkus, 2011; Killion & Harrison, 2006). The profession is just beginning to understand the role of mathematics coaches and the potential of these professionals to improve instruction and, in turn, student achievement.

We distinguish between *coaching* and *mentoring* by including coaching as a specific mentoring activity. Leading professional development, facilitating COLs, co-teaching, model teaching, and coaching are all considered mentoring, and each comprises different purposes and functions.

Characteristics of Good Mentors and Coaches

Many researchers in different content areas have attempted to document the characteristics of effective and ineffective coaches (e.g., Perkins, 1998). Ineffective mathematics coaches are unprofessional in front of their peers (e.g., grading papers during professional development) and lack passion for the teaching and learning field (Harrison, Higgins, Zollinger, Brosnan, & Erchick, 2011). In contrast, effective coaches share many of the characteristics of good COL facilitators. Kowal and Steiner (2007) suggest that coaches must have pedagogical knowledge, content knowledge, and interpersonal skills. Also key are a deep knowledge of the curriculum, coaching resources, and knowledge of coaching (Feger, Woleck, & Hickman, 2004) and the ability to establish a trusting, nonevaluative relationship and to ask generative rather than judgmental questions (Costa & Garmston, 1994).

The bottom line is that the people who are chosen for such an important role need to be trustworthy, knowledgeable about the content, and skilled in teaching the content.

We would argue additionally that a good mentor knows which mentoring strategy to use with which teacher and when. For example, Ms. Jewel may be the most experienced teacher for autonomy in the building, but she needs to improve her questioning techniques; a mentor might use a coaching approach with her. Ms. Nolsheim, in contrast, is a first-year teacher and has difficulty getting away from lecturing; a co-teaching or model-teaching strategy could be most effective for her. Hence, being able to analyze the needs of individual teachers as they travel along their personal learning routes is critical for a mentor.

We have argued elsewhere that teaching for autonomy requires highly specialized knowledge regarding a different type of planning and classroom teaching practices (Stephan, Underwood-Gregg, & Yackel, 2014). That is why we include experience in and knowledge of lesson imaging and teaching for autonomy as essential characteristics of both good mentors and good COL facilitators.

Coaching Programs

There are numerous types of coaching programs, some that are discipline-specific and others that are more general. *Cognitive coaching* (Costa & Garmston, 1994) is regarded as a productive approach that engages teachers in the deepest reflection on their practice. The ultimate goal of cognitive coaching is to teach teachers how to coach themselves by learning how to identify problematic areas in their practice, devise appropriate data collection procedures and analysis techniques, and determine where and how to change. The steps of the process are as follows:

1. The cognitive coach and teacher have a pre-conference; the teacher identifies a particular area of his or her practice that the teacher would like to explore more deeply, and together the teacher and coach create a data collection procedure.

2. The coach attends one of the teacher's classes and collects the data.

3. The coach and teacher have a post-conference, where the coach works with the teacher to analyze the data and make suggestions for improvement.

While we have found cognitive coaching to be the most sophisticated and compatible approach to use with teaching for autonomy, it is important to note that it is not always effective for all teachers. For example, after Dr. Stephan coached six different mathematics teachers at one school, she determined that first-year teachers were unable to participate in cognitive coaching because their practices were just developing and unstable. Their mentoring needs included creating an image of teaching for autonomy by watching the mentors teach in their classes and, at times, co-teaching with their mentor. Additionally, having the mentors observe their teaching and use a classroom practice rubric was incredibly informative for them. However, teachers who were midway toward forming their autonomy practices were able to engage very effectively in cognitive coaching.

Other effective coaching programs that might work better with certain teachers are *content-focused coaching* (West & Staub, 2003), *instructional coaching* (Kowal & Steiner, 2007), and *collaborative coaching and learning* (Neufeld, 2002). *Peer coaching* has been found to be the least effective approach (Murray, Ma, & Mazur, 2009). Regardless of the different mentoring techniques and coaching programs that are used, the administration must ensure that the person chosen as a mentor receives appropriate training on the most effective goals and approaches.

Mentoring
Chris Cline

An important aspect for a facilitator or mentor is to be a leader in the classroom. Mentors should be able to teach a class while others observe; they should be able to co-teach a particular lesson with another teacher and observe different teachers who are trying to implement this teaching approach. It is during these observation periods and discussions that mentors will show their greatest value. The coach and teacher can meet before the observation to discuss specifics about the class that is being observed. From personal experience, a mentor can and will ask the teacher more specific and meaningful questions about what is being observed. Mentors will ask questions that an administrator would not ask; they will ask questions of the teacher that are not part of the scripted questions that most administrators ask prior to observing a teacher.

The mentor's role should not be to tell a teacher what to do; it should be to guide a teacher to see areas where the teacher needs to get better. In some cases, the mentor should ask the observed teacher what the teacher wants the mentor to look for during the specific class. I, like many, wasn't able to answer this particular question when I participated in a coaching session with my mentor. It forced me to really think about my delivery of instruction. So I did what many others have done—I asked the coach, "What do you think?" The instructional facilitator who was coaching me forced me to think about my teaching practice and was able to guide me through a series of questions to identify a specific area of my practice that needed to be observed.

As Mr. Cline explains above, the coach was able to help him analyze his current practice to find a problematic area or at least an area of focus for improvement. Because he had more than 20 years of teaching experience at the point he participated in coaching, it was difficult for him to know what to focus on. During this coaching session, the coach, who happened to be Dr. Stephan, focused Mr. Cline on two specific areas: lesson imaging and whole-class discussion.

A Coaching Experience
Michelle Stephan

As a mentor for the staff of Chris Cline's middle school, I decided that Chris, with his 20 years of teaching experience, did not need to see me model-teach. He had already formed a pretty strong practice and was implementing lesson imaging on a consistent basis in his planning. Therefore, I thought that using a blend of cognitive and content-focused coaching would be the best course. I started with content-focused coaching, asking him content-focused questions about his lesson image. We then engaged in a cognitive coaching portion, where I let him choose what part of this new teaching approach he was struggling with most and would like to explore.

I will never forget asking him what his lesson objective was. He stated the objective as if he had copied it from a state standard. I asked him what it really meant for a student to master that particular objective. After exploring the answer to that question, he moved to the tasks he had selected for the lesson—and then realized that some of the tasks he had chosen did not lead to the objective for the lesson. It was such an incredible moment for me, to see his eyes light up right then. He found out that by unpacking the lesson objective, he was better able to choose problems for his students that would lead to his goal. I think it was an "aha" moment for him, too.

After anticipating different student solutions and working out the order that he would have students present, we moved on to the cognitive coaching part in which I asked him what data I could collect for him. It is here that he asked me what I thought. I would rather have the teacher self-identify a problematic area, but I also realized that, especially in the beginning, teachers may not know what types of problems to explore. So I mentioned that most teachers who are new to teaching for autonomy complain that they do not have productive whole-class discussions, despite having anticipated student thinking. I suggested that we focus on his monitoring techniques during small-group exploration. We agreed that I would follow him around the classroom during monitoring and write down the types of questions he asked students in their groups as well as which student solutions were shared with him. During whole-class discussion, I would record the students that he called on to present and in what order he chose them.

In our post-conference, Chris and I looked over the notes I had made, and he had another "aha" moment that he still talks about to this day. When he read through the interactions he was having with students, he learned that he was asking how students were thinking, but he was not stopping there. Rather than writing down students' diverse strategies to use in discussion, he was reacting to their strategies and, in essence, having mini whole-class discussions during explore time with each small group. When he moved to another table to check in, he did the same thing. The final debriefing discussion with the whole class was basically the same conversation that had happened in small groups, which made the discussion rather pointless.

Chris revised his monitoring technique for future class periods by planning to spend only two or three minutes at a group and to simply record their strategy on his paper. In this way, he was able to visit almost every group in 15 minutes and have a much fresher whole-class conversation that was led by students.

From this reflection, we see that the mentor analyzed the teacher's needs and decided which mentoring technique would be most productive for him. Dr. Stephan chose a combination of two coaching approaches with him, because she knew that the newness of lesson imaging probably necessitated some *content-focused coaching* (coaching that is specific to coaching the discipline, in this case, teaching mathematics for autonomy). Through certain questions about the lesson image process, Mr. Cline was able to prepare a stronger lesson that was more focused on the lesson objective. Because the mentor was skilled in lesson imaging, she knew to ask about unpacking the objective and making sure that it aligns with the chosen tasks. She then used another mentoring technique, cognitive coaching, to push Mr. Cline to examine an area of his practice that he was not even aware he might need to explore. Analyzing his data provided a chance for Mr. Cline to learn how to examine his own practice and make meaningful changes.

Mentors can make a difference in a teacher's practices, but it is not always as easy as our excerpt suggests. We cannot underscore the importance of choosing the right mentors. The mentoring experience above worked well because Mr. Cline and Dr. Stephan had established mutual trust and respect, and Dr. Stephan had experience with lesson imaging, knowledge of teaching for autonomy, pedagogical and content knowledge, and many of the other characteristics we have mentioned. However, not all of her mentoring experiences have been productive. Other teachers at the same school have never made shifts in their practice, even though their data illustrated a needed change. This brings us to the role of the administrator in supporting teacher change.

Administration

One more key area of support for lesson imaging that we'd like to discuss is the role that administrators play in initiating and sustaining change. Although it is more desirable for change to be motivated from the bottom up (i.e., by teachers), it is often dictated from the top down in terms of "the latest mandate from the district." Until that changes, administrators have an important role in facilitating the process of moving from standard practices to reform initiatives:

- Providing plenty of resources, both human and material
- Being willing to see a decline in scores at first, and staying the course despite unpopularity
- Analyzing data to inform administrative decisions
- Genuinely and productively communicating to all stakeholders (e.g., teachers, counselors, students, parents)

Probably the most important role the principal can play in supporting change is to provide the resources necessary for teachers to make the desired transition. Resources can come in at least two forms:

- Material resources, which include common planning time and ample classroom and curricular materials and laboratory equipment
- Human resources, which include STEM coaches, COL leaders, facilitators, substitute teachers, and administrators who understand the COL goals

---

Time as a Resource
Julie Cline

The first step in aiding our transition was for the administration to provide a common planning time. As mentioned earlier, this takes some thought, particularly if any COL members teach more than one subject. At our school, the entire grade level has the same planning period, which assists in making common time available. Our administration mandated that no other meetings were to be scheduled during that common COL time.

The principal also scheduled mandatory professional development meetings that focused on lesson imaging and had us read articles and books to aid in our growth. For example, a local professor joined our monthly department meeting to discuss our interpretation of the book *Five Practices for Orchestrating Productive Mathematics Discussions* (Smith & Stein, 2011), which covered some of the same practices that are involved in lesson imaging (anticipating student thinking, selecting and sequencing student explanations). We read a chapter or so each month and discussed them during a one-hour meeting.

The principal also mandated that all teachers in the school be a member of a COL that met at minimum once per week. While she did not dictate the content of those meetings for all disciplines, it was clear that the COLs in our department would be lesson imaging during that time. We might have other items on the agenda from time to time, but imaging is at the heart of every meeting.

Human Resources
Julie Cline

Administrators—both principals and assistant principals—should attend as many COL meetings as possible. This can be difficult due to lack of time, but if all members of the administrative team work together, it can be accomplished.

In previous years, this didn't happen in our meetings, and reluctant members often derailed the conversations and hindered the progress of imaging. Currently, an administrator attends each COL meeting, and the less-focused COL members are less apt to turn the discussion away from imaging.

Again, in the early stages of implementation, our COL had members that consistently did not do the prerequisite imaging for a sequence. The presence of an administrator has encouraged those members to come to meetings prepared. The administrators can also be very helpful when addressing non-normative behavior, such as COL members' negative dispositions, especially outside of meeting time.

The administration is likewise responsible for putting the right people in leadership roles, including assistant principals, coaches, and COL facilitators. We are fortunate to have math facilitators for two years, appointed by the administrative team. The team chose facilitators who were well versed in lesson imaging and could maintain a strong commitment to encouraging discourse and autonomy in math classrooms. What these facilitators have needed most from the administration is support when dealing with members who are reluctant to change their approach to instruction.

Principals are the leaders of their school in the community's eyes, so another important role in this process is to inform parents and students of the changes ahead. When parents complain that teachers are not telling their kids how to solve problems, the administration can support the teachers' work because they have been in the COL meetings and classrooms, they know how the content is being developed, and they understand how the classroom practices are changing.

One way to share the vision of the new approach with parents and their children is for STEM disciplines, individual departments, or a group of teachers to organize a STEM Family Fun Night. Family Fun Nights can be structured differently, but they should convey that the intent of teaching for autonomy is to prepare children to participate effectively in the digital age.

• One middle school mathematics department invited all parents and their children to a Family Fun Night in which each family went to a smaller classroom and engaged in a STEM activity the same way that their children do each day. The parents watched their children invent their own solutions to problems and present their thinking in front of a group of unknown adults and students, thus recognizing the powerful approach being used in the classrooms. The principal organized the event, provided money and class-room resources to feed the parents, and was present to introduce the teachers and provide verbal support.

• Another teacher held a Family Fun Night with just her own students and their parents. These parents also engaged in a class one evening, where they were taught a lesson that their children had already participated in. The same mathematical discussions occurred with the parents and allowed them to see how students were gaining autonomy in their own mathematical thinking. Prior to this night, many parents had complained that teachers were not helping students. The teacher was able to address the benefits of *productive struggle* versus simply "not helping" students.

Professional Development for Administrators
Michelle Stephan

In an ideal world, resources would be used to train all administrators, including counselors, to understand teaching for autonomy. In my first year of teaching middle school, students lined the guidance counselors' hallway in order to get transferred out of my class. Luckily, the counselors understood my goals and knew how to communicate with students and parents about the goal of the class. Additionally, when administrators visited my classroom for formal or informal observations, they knew what to look for, and they expected to hear what might sound like chaotic conversation. Administrators with little training might interpret the same classroom as "loud" or disorganized and give low marks to the teacher.

Administrators must be dedicated to the change in methodology and ready to stay the course despite possibly having a dip in test scores. Like with any other changes, the long-term, systemwide results are not going to become evident overnight. On the other hand, staying the course after data repeatedly show a decline is not prudent either. Thus, administrators, like their teachers, must use data to inform their administrative decisions.

For example, the first year that Kristi Bullock, one of our colleagues, implemented such a change with her staff, she used end-of-the-year data to make decisions about the progress (not the termination) of the program. When she dug deeply into students' test scores, she noticed that the COL that was most resistant to change and overtly ignored the mandate had a huge drop in test scores. The COL that lesson imaged the most and was dedicated to making change had the highest growth data of the seven middle schools in the district. This led her to make important decisions about which grade teachers would teach the following year as well as what subject.

We have seen principals move teachers out of the STEM discipline to teach another subject for which they are certified. This is a move that Collins (2001) would call getting people in the right seat on the bus.

Conclusion

In this chapter, we presented strategies to help teachers in the process of learning to lesson image. We also outlined resources (common planning time, instructional materials that promote autonomy, mentoring, specific professional development, COLs) and practices (lesson imaging, attendance at COL meetings) that principals can provide to facilitate teachers' transition to teaching for autonomy.

Initiating and sustaining a major change initiative at the school level is a complex process. While it is best when change is introduced from the bottom up, from teachers who see the need, this is not always possible. In cases where the catalyst comes from someone other than the teachers, it is best to begin by helping teachers see the need for that change. Principals can be instrumental in sparking that desire by creating a sense of urgency among their faculty. In every school we have worked with, that sense of urgency was created by working with teachers to understand their students' low scores on state tests. Traditional instruction with lecture-oriented classroom materials was not satisfying the needs of their students, so teachers were more willing to entertain change from without.

Hopefully, this chapter has made it clear that classroom change is not an individual journey but, rather, a coordinated effort of a number of stakeholders. Teachers cannot travel this path alone; they must rely on other teachers to share their expertise in the lesson imaging process. Teachers also depend on the resources provided by the administration, including time, knowledgeable mentors, verbal support to students and parents, and a willingness to let teachers experiment in their classrooms. Even then, change takes time.

Frequently Asked Questions

Below, we answer nine questions that have been asked of us repeatedly.

I am a principal, and I have a teacher who will not get on board. What do I do?

Kristi Bullock, middle school STEM principal, Concord, North Carolina: It is not uncommon to encounter teachers who are reluctant to transition to this approach to instruction. Here are my 10 tips for facilitating change from an administrator's perspective:

1. Do not allow teachers to make excuses.
2. Give teachers permission to fail if they are making a genuine effort.
3. Encourage risk taking.
4. Provide resources such as common planning time, less-traditional materials, and coaches.
5. Articulate a long-term implementation plan with short-term goals and ongoing evaluation.
6. Give teachers permission to veer from the district-mandated curriculum documents when justified.
7. Provide financially supported summer professional development to image and plan long range.
8. Raise administration's expectations of teachers for the quality of planning time to include lesson imaging.

9. Be physically present in teachers' lesson imaging sessions and classrooms to show administrative support.

10. Hold teachers accountable for lesson imaging.

What supports do you think teachers and coaches need to be successful?

Kristi Bullock, middle school STEM principal, Concord, North Carolina: In order to support your teachers, the entire administration team needs to know what lesson imaging is and what good STEM instruction looks like when they enter a classroom, and have a clear articulation between teachers and administration about what is expected of them. Teachers need to have time to lesson image, even if it requires hiring a half-day substitute for the community of learners (COL) to meet occasionally. Common planning time is incredibly important and can usually be arranged despite some complaints that it is too difficult to schedule. Instructional coaches are imperative in order to model or co-teach this approach, observe, and give feedback to growing teachers. These instructional coaches should have limited, if any, classroom instructional responsibilities and should not be asked to fill in for the teachers who are absent from work. The coaches' time should be spent with teachers in their classrooms. Additionally, the instructional coaches need coaching and professional development themselves in order to improve their teacher support.

I am a facilitator, and I have a teacher who will not get on board. What do I do?

Julie Cline, middle school teacher, Concord, North Carolina: This is a tough one! When this first happened to me, I tried to get to the root of why a teacher wasn't on board. I was empathic to teachers who resisted due to little faith in the methodology. I would have a very difficult time teaching in a way I didn't believe in! In my experience, the biggest resistance came from teachers who claimed that they already taught for autonomy but really didn't. The need for change was not obvious to them. This is a situation that requires coaching, collecting data, and reflection.

My questions for the resistant teacher would be, "What do you believe is best practice for students in your classroom? What should it look and sound like? How do you plan to create this ideal classroom?" I would then have the teacher observe some classrooms that use imaging and reflect on how those classrooms align with the teacher's vision.

However, even with these strategies and the help of your administration, there may come a point where you will have to accept that a teacher will not change. Do not let one teacher derail the advancement of the COL. Continue to use your protected COL time to lesson image.

I am a coach, and one of my teachers thinks she doesn't need help. What strategies can I use to help her reflect on her practice?

Julie Cline, middle school teacher, Concord, North Carolina: I have found that collecting data for teachers is a great tool for reflection. I typically start by asking what a teacher thinks he or she does well, then follow up with an area where the teacher thinks he or she could improve. In most cases, teachers can come up with something on their own. If not, I share some areas that tend to be difficult for me and that overlap with lesson imaging practices. We then brainstorm together how I can collect data to reflect that piece of the teacher's practice. I schedule some time in the classroom to collect the data and then follow up to share the data. Colleagues are more receptive to receiving the data we agreed to collect than simply receiving a critique from me.

My school does not have common planning time. How can we lesson image?

Julie Cline, middle school teacher, Concord, North Carolina: In this case, it may be necessary for your COL to meet before or after school. Another suggestion is that teachers across grade levels work to cover each other's classes once a week so that a common planning time can be scheduled.

I am the only teacher who wants to teach for autonomy. How do I do this alone?

Julie Cline, middle school teacher, Concord, North Carolina: For the teacher who is just beginning this process and is alone, using the lesson imaging template is vital:

> • The template outlines the important pieces to implement this new approach successfully, such as anticipating student responses.
> • It gives users a place where they can select and sequence different student responses.
> • It keeps the mathematical goal of the lesson and the connection to the standards at the forefront.

As you become more experienced in the inquiry process, you might find that you want to keep your imaging notes in other places. Some people I have worked with keep their imaging notes in separate notebooks, while others keep their notes in the margins of their curriculum.

Is this approach to teaching and imaging appropriate for students with mild disabilities?

Jennifer Smith, Special Education Department chair, Florida: As a teacher of students with disabilities, I have found that lesson imaging is a way for me to maximize all students' learning, but in particular those with disabilities. This teaching and planning approach changed my instruction significantly and also my mindset of teaching students with disabilities. Changing the way I teach from telling the students how to come up with the solution (i.e., 100 percent direct instruction) to allowing the students to discover the solutions with limited guidance (teaching for autonomy) made all the difference to my students' mathematical learning and, ultimately, test scores. I will never forget the day that one of my students said to me, "Ms. Smith, no one has ever asked me how I thought about a problem before you." I realized then that my direct instruction forced my way of doing mathematics on him and that I was not taking advantage of the "out of the box" thinking that many of my students use when problem solving.

How does the special educator find the time to lesson image?

Erika Allred, Special Education Department chair, North Carolina: I think the most important thing is to make sure that the teachers have protected time to work together and that the special education teacher has the same time to work with the co-teacher. The most important thing for a co-teaching pair is to have the same co-teacher from year to year. This is so important because each teacher needs to know how the other thinks in order for this approach to be the most successful, and the only way for that to happen is to allow them to work together, year after year.

How does differentiation work in lesson imaging?

Erika Allred, Special Education Department chair, North Carolina: As a special educator, I believe that this approach to teaching makes differentiation in the lesson much easier and less invasive for the special education students. It doesn't single them out or make them feel any different from their peers. In this approach, you are already differentiating the students' learning by allowing them to choose which option best fits their learning style.

When imaging a lesson, the most important thing to remember about differentiation is that the simplest but longest way to solve the problem may be the way that a student with disabilities chooses to solve all problems, and that is OK. This manner of teaching has allowed many of my students with disabilities to feel successful in math for the first time in their lives. They can explain their thinking to another student through pictures, charts, and many other methods that my co-teacher or I may never come up with, but these methods "click" for them.

References

Akyuz, D., Stephan, M., & Dixon, J. K. (2013). Improving the quality of mathematics teaching with effective planning practices. *Teacher Development, 17*(1), 92–106.

Battista, M. (2004). Applying cognition-based assessment to elementary school students' development of understanding of area and volume measurement. *Mathematical Thinking and Learning, 6*(2), 185–204. doi:10.1207/s15327833mtl0602_6

Battista, M. T., & Clements, D. H. (1996). Students' understanding of three-dimensional rectangular arrays of cubes. *Journal for Research in Mathematics Education, 27*(3), 258–292.

Binkley, M., Erstad, O., Herman, J., Raizen, S., Ripley, M., Miller-Ricci, M., & Rumble, M. (2012). Defining twenty-first century skills. In P. Griffin, B. McGaw, & E. Care (Eds.), *Assessment and teaching of 21st century skills* (pp. 17–66). Dordrecht, the Netherlands: Springer.

Bloom, B. S. (Ed.). (1972). *Taxonomy of educational objectives book 1—Cognitive domain.* New York: David McKay Company.

Campbell, P. F., & Malkus, N. N. (2010, May). *The impact of elementary mathematics coaches on teachers' beliefs and professional activity.* Paper presented at the annual meeting of the American Educational Research Association, Denver, CO.

Campbell, P. F., & Malkus, N. N. (2011). The impact of elementary mathematics coaches on student achievement. *The Elementary School Journal, 111*(3), 430–454.

Cheuk, T. (2012). *Relationships and convergences among the mathematics, science, and ELA practices.* Palo Alto, CA: Stanford University. Retrieved from http://ell.stanford.edu/content/science

Clarke, D. (2014). *Disciplinary inclusivity in educational research design: Permeability and affordances in STEM education.* Retrieved from http://stem2014.sites.olt.ubc.ca/files/2014/07/Permeability-and-Affordances-in-STEM.pdf

Cobb, P., & McClain, K. (2006). The collective mediation of a high-stakes accountability program: Communities of networks and practices. *Mind, Culture and Activity, 13*(2), 80–100.

Cobb, P., Yackel, E., & Wood, T. (1989). Young children's emotional acts while engaged in mathematical problem solving. In D. B. McLeod & V. M. Adams (Eds.), *Affect and mathematical problem solving* (pp. 117–148). New York: Springer Verlag New York.

Collins, J. C. (2001). *Good to great: Why some companies make the leap… and others don't.* New York: HarperBusiness.

Committee on Standards for K–12 Engineering Education & National Research Council. (2010). *Standards for K–12 Engineering Education.* Washington, DC: The National Academies Press.

Costa, A. L., & Garmston, R. J. (1994). *Cognitive coaching: A foundation for renaissance schools.* Norwood, MA: Christopher-Gordon.

Davis, B. (1997). Listening for differences: An evolving conception of mathematics teaching. *Journal for Research in Mathematics Education, 28*(3), 355–376.

Delors, J. (2013). The treasure within: Learning to know, learning to do, learning to live together and learning to be. What is the value of that treasure 15 years after its publication? *International Review of Education, 59*(3), 319–330.

Dieker, L. (2001). What are the characteristics of "effective" middle and high school co-taught teams for students with disabilities? *Preventing School Failure, 46*(1), 14–23.

DuFour, R. (2004). What is a professional learning community? *Educational Leadership, 61*(8), 6–11.

English, L. (2015). STEM: Challenges and opportunities for mathematics learning. In K. Beswick, T. Muir, & J. Wells (Eds.), *Proceedings of the 39th Conference of the International Group for the Psychology of Mathematics Education* (Vol. 1, pp. 3–18). Hobart, Australia: International Group for the Psychology of Mathematics Education.

Evagorou, M., Erduran, S., & Mäntylä, T. (2015). The role of visual representations in scientific practices: From conceptual understanding and knowledge generation to "seeing" how science works. *International Journal of STEM Education, 2*(1), 1–13.

Feger, S., Woleck, K., & Hickman, P. (2004). How to develop a coaching eye. *JSD: The Learning Forward Journal, 25*(2), 14–18.

Fuchs, D., Fuchs, L., & Burish, P. (2000). Peer-assisted learning strategies: An evidence-based practice to promote reading achievement. *Learning Disabilities Research and Practice, 15*(2), 85–91.

Harrison, R., Higgins, C., Zollinger, S., Brosnan, P., & Erchick, D. (2011). Mathematics Coaching Program fidelity and its impact on student achievement. Brief research report. In L. R. Weist & T. Lamberg (Eds.), *Proceedings of the 33rd Annual Meeting of the North American Chapter of the International Group for the Psychology of Mathematics Education* (pp. 1545–1553). Reno, NV: University of Nevada, Reno.

Herbel-Eisenmann, B., & Breyfogle, M. (2005). Questioning our patterns of questioning. *Mathematics Teaching in the Middle School, 10*(9), 484–489.

Hurd, P. D. (1958). Science literacy: Its meaning for American schools. *Educational Leadership, 16*, 13–16, 52.

Inoue, N., & Buczynski, S. (2011). You asked open-ended questions, now what? Understanding the nature of stumbling blocks in teaching inquiry lessons. *The Mathematics Educator, 20*(2), 10–23.

International Technology Education Association. (2007). *Standards for technological literacy: Content for the study of technology* (3rd ed.). Reston, VA: Author. Retrieved from https://www.iteea.org/File.aspx?id=67767&v=691d2353

Jackson, K., Shahan, E., Gibbons, L., & Cobb, P. (2012). Launching complex tasks. *Mathematics Teaching in the Middle School, 18*(1), 24–29.

Kamii, C. (1982). *Number in preschool and kindergarten.* Washington, DC: National Association for the Education of Young Children.

Karahan, E., Guzey, S., & Moore, T. (2014). Saving pelicans: A STEM integration unit. *Science Scope, 38*(3), 28–34.

Killion, J., & Harrison, C. (2006). *Taking the lead: New roles for teachers and school-based coaches.* Oxford, OH: National Staff Development Council.

Kowal, J., & Steiner, L. (2007, September). *Instructional coaching.* The Center for Comprehensive School Reform and Improvement. Issue Brief (1–8).

Lappan, G., Phillips, E. D., Fey, J. T., & Friel, S. N. (2013). *Connected Mathematics 3. Thinking with mathematical models: Linear and inverse variation.* Boston: Pearson.

Mack, N. (2001). Building on informal knowledge through instruction in a complex content domain: Partitioning, unit, and understanding multiplication of fractions. *Journal for Research in Mathematics Education, 32*(3), 267–295.

Mathematics Learning Study Committee. (2001). *Adding it up: Helping children learn mathematics.* Washington, DC: The National Academies Press.

Mehan, H. (1979). "What time is it, Denise?" Asking known information questions in classroom discourse. *Theory into Practice, 18*(4), 285–294.

Murray, S., Ma, X., & Mazur, J. (2009). Effects of peer coaching on teachers' collaborative interactions and students' mathematics achievement. *Journal of Educational Research, 102*(3), 203–212.

Nan-Zhao, Z. (2008). *Four "pillars of learning" for the reorientation and reorganization of curriculum: Reflections and discussions.* Paris: UNESCO. Retrieved from https://www.ibe.unesco.org/cops/Competencies/PillarsLearningZhou.pdf

National Council of Teachers of Mathematics. (2000). *Principles and standards for school mathematics.* Reston, VA: Author.

National Governors Association Center for Best Practices & Council of Chief State School Officers. (2010). *Common Core State Standards.* Washington, DC: Authors.

Neufeld, B. (2002). *Using what we know: Implications for scaling up implementation of the CCL model.* Cambridge, MA: Education Matters, Inc.

NGSS Lead States. (2013). *Next Generation Science Standards: For states, by states.* Washington, DC: The National Academies Press.

Organisation for Economic Co-operation and Development. (2007). *PISA 2006: Science competencies for tomorrow's world: Volume 1: Analysis.* Paris: Author. doi:10.1787/9789264040014-en

Perkins, S. J. (1998). On becoming a peer coach: Practices, identities, and beliefs of inexperienced coaches. *Journal of Curriculum and Supervision, 13*(3), 235–254.

Piaget, J. (1948/1973). *To understand is to invent.* New York: Grossman.

Polya, G. (1957). *How to solve it: A new aspect of mathematical method* (2nd ed.). Princeton, NJ: Princeton University Press.

Prud'homme-Généreux, A. (2011). *The evolution of human skin color.* Buffalo, NY: National Center for Case Study Teaching in Science, University at Buffalo, State University of New York. Retrieved from http://sciencecases.lib.buffalo.edu/cs/files/skin_pigmentation.pdf

Puttnam, D. (2015). Learning to be—Do our education systems do enough to enable learners to flourish as independent, autonomous and well-balanced individuals? *European Journal of Education, 50*(2), 120–122.

Quarteroni, A. (2009). Mathematical models in science and engineering. *Notices of the American Mathematics Society, 56*(1), 10–19.

Romberg, T. A., & de Lange, J. (1998). *Mathematics in context.* Chicago: Encyclopaedia Britannica.

Schoenfeld, A. H. (1998.) Toward a theory of teaching-in-context. *Issues in Education, 4*(1), 1–94.

Smith, M., & Stein, M. (2011). *Five practices for orchestrating productive mathematics discussions.* Reston, VA: National Council of Teachers of Mathematics.

Stage, E. K., Asturias, H., Cheuk, T., Daro, P. A., & Hampton, S. B. (2013). Opportunities and challenges in Next Generation Standards. *Science, 340*(6130), 276–277.

Steen, L. A. (2003). Data, shapes, symbols: Achieving balance in school mathematics. In B. L. Madison & L. A. Steen (Eds.), *Quantitative literacy: Why numeracy matters for schools and colleges* (pp. 53–74). Princeton, NJ: National Council on Education and the Disciplines.

Stephan, M., Akyuz, D., McManus, G., & Smith, J. (2012). Conditions that support the creation of mathematical communities of teacher learners. *NCSM Journal of Mathematics Education Leadership, 14*(1), 19–27.

Stephan, M., & McManus, G. (2013). *Pre-interviewing as formative assessment.* Paper presented at the Bi-Annual International Symposium in Elementary Mathematics Teaching, Prague, Czech Republic.

Stephan, M., McManus, G., & Dehlinger, R. (2014). Using research to inform formative assessment techniques. In K. Karp (Ed.), *NCTM annual perspectives in mathematics education: Using research to improve instruction* (pp. 229–238). Reston, VA: NCTM.

Stephan, M., McManus, G., Smith, J., & Dickey, A. (n.d.). *Ratio and rates.* Retrieved from http://cstem.uncc.edu/sites/cstem.uncc.edu/files/media/Ratio T Manual.pdf

Stephan, M., Underwood-Gregg, D., & Yackel, E. (2014). Guided reinvention: What is it and how do teachers learn this teaching approach? In Y. Li (Ed.), *Transforming mathematics instruction: Multiple approaches and practices, Advances in Mathematics Education* (pp. 37–57). Cham, Switzerland: Springer International Publishing.

Stephan, M., & Whitenack, J. (2003). Establishing classroom social and sociomathematical norms for problem solving. In F. Lester (Ed.), *Teaching mathematics through problem solving: Prekindergarten–grade 6* (pp. 149–162). Reston, VA: National Council of Teachers of Mathematics.

Stevenson, H. W., & Stigler, J. W. (1992). *The learning gap: Why our schools are failing, and what we can learn from Japanese and Chinese education.* New York: Summit Books.

Thompson, A., Philipp, R., Thompson, P., & Boyd, B. (1994). Calculational and conceptual orientations in teaching mathematics. In *1994 yearbook of the National Council of Teachers of Mathematics* (pp. 79–92). Reston, VA: National Council of Teachers of Mathematics.

Travers, K. J. (1993). Overview of the longitudinal version of the Second International Mathematics Study. In L. Burstein (Ed.), *The IEA study of mathematics III: Student growth and classroom processes* (pp. 1–27). New York: Pergamon.

Underwood, D. (2002). Building students' sense of linear relationships by stacking cubes. *Mathematics Teacher, 95*(5), 330–333.

Underwood, D., & Yackel, E. (2002). Helping students make sense of algebraic expressions: The candy shop. *Mathematics Teaching in the Middle School, 7*(9), 492–497.

Voigt, J. (1995). Thematic patterns of interaction and sociomathematical norms. In P. Cobb & H. Bauersfeld (Eds.), *The emergence of mathematical meaning: Interaction in classroom cultures* (pp. 163–201). Hillsdale, NJ: Lawrence Erlbaum.

Warshauer, H. K. (2015). Strategies to support productive struggle. *Mathematics Teaching in the Middle School, 20*(7), 390–393.

Welborn, J. (2013). 99.99%: Antibacterial products and natural selection. *Science Scope, 37*(4), 58–68.

West, L., & Staub, F. C. (2003). *Content-focused coaching: Transforming mathematics lessons.* Portsmouth, NH: Heinemann; and Pittsburgh, PA: University of Pittsburgh.

Wiggins, G., & McTighe, J. (2005). *Understanding by design* (2nd ed.). Alexandria, VA: ASCD.

Yackel, E., & Cobb, P. (1996, July). Sociomathematical norms, argumentation, and autonomy in mathematics. *Journal for Research in Mathematics Education, 27*(4), 458–477.

Zollman, A. (2012). Learning for STEM literacy: STEM literacy for learning. *School Science and Mathematics, 112*(1), 12–19.

Index

The letter *f* following a page number denotes a figure.

About the Authors

Michelle Stephan, EdD, is an associate professor of mathematics education with a joint appointment in the College of Education and Department of Mathematics and Statistics at the University of North Carolina at Charlotte. She earned bachelor's and master's degrees in pure mathematics and in 1998 earned an EdD in mathematics education from Peabody College of Education at Vanderbilt University. After working as an associate professor at Purdue University Calumet, she left academia for the classroom, where she taught middle school mathematics for seven years in Florida. It was there that she began to explore the ideas of lesson imaging with her middle school teaching colleagues George McManus, Jennifer Smith, and Ashley Dickey. In 2012, Stephan became faculty at the University of North Carolina at Charlotte and began a collaboration with coauthors David Pugalee, Julie Cline, and Chris Cline. Together they have presented on lesson imaging at numerous conferences and provided professional development for several North Carolina school districts. Stephan has published a dozen book chapters and more than 25 journal articles, and she has made more than 50 presentations both nationally and internationally.

David Pugalee, PhD, is a professor of education at the University of North Carolina at Charlotte, where he serves as director of the Center for Science, Technology, Engineering, and Mathematics Education. He earned his PhD in mathematics education from the University of North Carolina at Chapel

Hill and has also taught at the elementary, middle, and secondary levels. With more than a decade of experience teaching mathematics and science, he has published research articles in *American Educational Research Journal, Educational Studies in Mathematics,* and *School Science and Mathematics.* His works include several books and book chapters published by the National Council of Teachers of Mathematics. In addition, Pugalee has published two books on communication and mathematical and scientific literacy: *Writing to Develop Mathematical Understanding* and *Effective Content Reading Strategies to Develop Mathematical and Scientific Literacy.* His research interest is the relationship between language and mathematics teaching and learning.

Julie Cline earned a bachelor's degree in education from Wingate University in North Carolina. She has taught middle school mathematics for 22 years and recently earned National Board Certification in Mathematics/Early Adolescence. After being introduced to Michelle Stephan, Cline began exploring lesson imaging in her practice. As the leader of her professional learning community, Cline encouraged other teachers to join her in using lesson imaging. She has provided professional development in her district as well as presented at several conferences. Cline continues to support teachers at her school in her current role as math facilitator.

Chris Cline is a graduate of the University of North Carolina at Charlotte, and holds a Bachelor of Arts degree in mathematics and a master's degree in mathematics education. He is also a National Board–Certified Teacher in Mathematics/Early Adolescence. Cline has taught 7th and 8th grade mathematics for 21 years. Three years ago, he incorporated lesson imaging into his practice and since has provided professional development in his district and has presented at several conferences. Cline has served as math facilitator at his school and has supported and coached colleagues.

Related ASCD Resources

At the time of publication, the following ASCD resources were available (ASCD stock numbers in parentheses). For up-to-date information about ASCD resources, go to www.ascd.org. Search the complete archives of *Educational Leadership* at www.ascd.org/el.

ASCD Edge®

Exchange ideas and connect with other educators interested in math and science on the social networking site ASCD Edge® at http://edge.ascd.org.

Print Products

Engaging Minds in Science and Math Classrooms: The Surprising Power of Joy by Eric Brunsell and Michelle A. Fleming (#113023)

Level Up Your Classroom: The Quest to Gamify Your Lessons and Engage Your Students by Jonathan Cassie (#116007)

Succeeding with Inquiry in Science and Math Classrooms by Jeff C. Marshall (#113008)

Real-World Projects: How do I design relevant and engaging learning experiences? (ASCD Arias) by Suzie Boss (#SF115043)

Ditch the Daily Lesson Plan: How do I plan for meaningful student learning? (ASCD Arias) by Michael Fisher (#SF116036)

Engineering Essentials for STEM Instruction: How do I infuse real-world problem solving into science, technology, and math? (ASCD Arias) by Pamela Truesdell (#SF114048)

DVDs

The Innovators: STEM Your School (#613042)

Meaningful Mathematics: Leading Students Toward Understanding and Application (#607085)

PD Online® Courses

STEM for All (#PD14OC012M)

For more information: send e-mail to member@ascd.org; call 1-800-933-2723 or 703-578-9600, press 2; send a fax to 703-575-5400; or write to Information Services, ASCD, 1703 N. Beauregard St., Alexandria, VA 22311-1714 USA.